ANODIC OXIDATION
OF ALUMINIUM
AND ITS ALLOYS

THE PERGAMON MATERIALS ENGINEERING PRACTICE SERIES
Editorial Board
Chairman: D. W. HOPKINS, University College of Swansea
J. R. BARRATT, British Steel Corporation
T. BELL, University of Birmingham
G. E. SHEWARD, UKAEA, Springfields Laboratories
A. J. SMITH
J. R. THORNTON, Iron and Steel Industry Training Board
Secretary: A. POST

OTHER TITLES IN THE SERIES

ALLSOP & KENNEDY	Pressure Diecasting, Part II
BAKER	Introduction to Aluminium Alloys
DAWSON	Welding of Non Ferrous Metals
LANSDOWN	Lubrication
PARRISH & HARPER	Production Gas Carburising
UPTON	Pressure Diecasting, Part I

NOTICE TO READERS

Dear Reader

An Invitation to Publish in and Recommend the Placing of a Standing Order to Volumes Published in this Valuable Series

If your library is not already a standing/continuation order customer to this series, may we recommend that you place a standing/continuation order to receive immediately upon publication all new volumes. Should you find that these volumes no longer serve your needs, your order can be cancelled at any time without notice.
The Editors and the Publisher will be glad to receive suggestions or outlines of suitable titles, reviews or symposia for editorial consideration: if found acceptable, rapid publication is guaranteed.

ROBERT MAXWELL
Publisher at Pergamon Press

ANODIC OXIDATION OF ALUMINIUM AND ITS ALLOYS

V F HENLEY
B.Sc.C.Chem., FRSC, C.Eng.FIM, FIMF
Consultant

PERGAMON PRESS

OXFORD · NEW YORK · TORONTO · SYDNEY · PARIS · FRANKFURT

U.K.	Pergamon Press Ltd., Headington Hill Hall, Oxford OX3 0BW, England
U.S.A.	Pergamon Press Inc., Maxwell House, Fairview Park, Elmsford, New York 10523, U.S.A.
CANADA	Pergamon Press Canada Ltd., Suite 104, 150 Consumers Road, Willowdale, Ontario M2J 1P9, Canada
AUSTRALIA	Pergamon Press (Aust.) Pty. Ltd., P.O. Box 544, Potts Point, N.S.W. 2011, Australia
FRANCE	Pergamon Press SARL, 24 rue des Ecoles, 75240 Paris, Cedex 05, France
FEDERAL REPUBLIC OF GERMANY	Pergamon Press GmbH, 6242 Kronberg-Taunus, Hammerweg 6, Federal Republic of Germany

Copyright © 1982 Pergamon Press Ltd.

All Rights Reserved. No part of this publication may be reproduced, stored in a retrieval system or transmitted in any form or by any means: electronic, electrostatic, magnetic tape, mechanical, photocopying, recording or otherwise, without permission in writing from the publishers

First edition 1982

Library of Congress Cataloging in Publication Data
Henley, V. F.
Anodic oxidation of aluminium and its alloys.
(Materials engineering practice)
Bibliography: p.
Includes index.
1. Aluminium—Anodic oxidation. 2. Aluminium alloys—Anodic oxidation. I. Title. II. Series
TS694.2.H46   1981   673'.7227   81-12039

British Library Cataloguing in Publication Data
Henley, V. F.
Anodic oxidation of aluminium and its alloys.
- (Materials engineering practice)
1. Aluminium-Anodic oxidation
2. Aluminium alloys - Anodic oxidation
I. Title II. Series
673'.722732   TS694.2
ISBN 0-08-026726-2
ISBN 0-08-026725-4 pbk

Printed in Great Britain by A. Wheaton & Co. Ltd., Exeter

Materials Engineering Practice

FOREWORD

The title of this new series of books "Materials Engineering Practice" is well chosen since it brings to our attention that in an era where science, technology and engineering condition our material standards of living, the effectiveness of practical skills in translating concepts and designs from the imagination or drawing board to commercial reality is the ultimate test by which an industrial economy succeeds.

The economic wealth of this country is based principally upon the transformation and manipulation of *materials* through *engineering practice*. Every material, metals and their alloys and the vast range of ceramics and polymers, has characteristics which require specialist knowledge to get the best out of them in practice, and this series is intended to offer a distillation of the best practices based on increasing understanding of the subtleties of material properties and behaviour and on improving experience internationally. Thus the series covers or will cover such diverse areas of practical interest as surface treatments, joining methods, process practices, inspection techniques and many other features concerned with materials engineering.

It is to be hoped that the reader will use this book as the base on which to develop his own excellence and perhaps his own practices as a result of his experience and that these personal developments will find their way into later editions for future readers. In past years it may well have been true that if a man made a better mousetrap the world would beat a path to his door. Today however to make a better mousetrap requires more direct communication between those who know how to make the better mousetrap and those who wish to know. Hopefully this series will make its contribution towards improving these exchanges.

MONTY FINNISTON

Preface

The new series of books on aluminium, to be published by Pergamon Press in its Materials Engineering Practice series, will fill the gap left when the Aluminium Federation ceased publishing its technical information bulletins. They will be useful not only to students and technicians in the aluminium industry, but also to product designers, architects, and engineers in other fields who need to work with aluminium.

The first in the aluminium series to be published, *Anodic Oxidation of Aluminium and its Alloys,* is based on one of the original bulletins, but its distinguished author has used his own knowledge and experience in bringing it up to date, extending it, and improving it. Best of all, Vernon Henley has included the numerous practical tips that spring from fifty years in the industry.

In those fifty years Mr. Henley has contributed a lot to the development of anodizing, both as Technical Director of Acorn Anodising Co. Ltd. and as Chairman of British Standards Institution committees on anodizing. He has also led British delegations dealing with anodizing in the International Standards Organisation. Mr. Henley founded the Anodizing Group—later the Aluminium Finishing Group—in the Institute of Metal Finishing.

Although his book is not a formal textbook, and is aimed more at the user of aluminium than at anodizing technicians, it covers the whole field very thoroughly, from the basic principles to choice of materials, pretreatment, design, properties of the anodic film, testing, and maintenance. The book is completely self-contained, but its value will no doubt be enhanced by other books in the series when they are published.

In wishing Pergamon every success with the series, one can only hope that the other books will be as good as this one.

1982

RICHARD WILTSHIRE
ALUMINIUM FEDERATION

Acknowledgements

Due acknowledgement has been given to all those who supplied new illustrations and data for this book. Tables and illustrations from the Aluminium Federation Bulletin No. 14 carry the original acknowledgements.

My special thanks are due to the Aluminium Federation for their encouragement and to their Dick Wiltshire for his many helpful suggestions.

Contents

INTRODUCTION			1
CHAPTER	1	PRINCIPLES OF ANODIZING	3
	2	APPLICATIONS OF ANODIZED ALUMINIUM	6
	3	FACTORS INFLUENCING THE CHOICE OF GRADE OF ALUMINIUM FOR ANODIZING	11
	4	FACTORS INFLUENCING THE CHOICE OF ANODIZING PROCESS	17
	5	DESIGNING FOR ANODIZING	23
	6	ANODIZING EQUIPMENT	29
	7	JIGGING (RACKING) METHODS FOR ANODIZING	42
	8	CHEMICAL TREATMENT PROCESSES BEFORE ANODIZING	49
	9	ANODIZING PROCESSES	56
	10	COLOURING THE ANODIC COATING	76
	11	SEALING THE ANODIC COATING	88
	12	STRIPPING ANODIC COATINGS AND DYES	97
	13	TESTING ANODIZED ALUMINIUM	100
	14	THE PROPERTIES OF ANODIZED ALUMINIUM	124
	15	MAINTENANCE OF ANODIZED ALUMINIUM	141
	16	EFFLUENT TREATMENT FOR ANODIZING PLANTS	143
APPENDIX	I	APPROXIMATE EQUIVALENT CONCENTRATIONS OF SULPHURIC ACID IN DIFFERENT UNITS	150

II	BATH ANALYSIS METHODS	152
III	SELECTED BOOKS—INFORMATION SOURCES	161
IV	SPECIFICATIONS APPLICABLE TO ANODIC OXIDE COATINGS ON ALUMINIUM	163

INDEX 167

Introduction

Aluminium and aluminium alloys* have some inherent resistance to atmospheric corrosion due to the presence of a protective oxide film that forms immediately the metal is exposed to air. This oxide film is about 0.1-0.4×10^{-6} in. or 0.25-1×10^{-2} microns† thick. Anodic oxidation, or anodizing, is an electrolytic process for producing very much thicker oxide coatings whose improved physical and chemical properties have greatly increased the field of application for aluminium.

The anodic oxide coating, when properly produced, has excellent resistance to marine and general atmospheric corrosion, is abrasion resistant, an electrical insulator and absorbs dyestuffs to give a wide range of colours.

On suitable material bright transparent coatings can be formed for decorative or optical use.

Some anodizing processes give coloured coatings, varying from pale yellow through to bronze and black.

The anodizing industry is firmly established in all the industrial countries and many of the emerging nations have also adopted anodizing as a finishing process, often in conjunction with the production of semi-fabricated products, particularly extruded sections.

The required properties and test methods for anodic oxide coatings are the subject of many national standards which are now being replaced or supplemented by International Standards.

The anodizing process is usually applied after any forming or machining operations, but it is commercially possible to produce relatively thin coatings that will withstand mild forming. In the

---

*In this book the word "aluminium" includes aluminium alloys unless specifically stated otherwise.

†The micron (micrometre)(μm) is one-millionth of a metre, 0.001mm (0.00004 in. approx.), and is widely used to describe the thickness of oxide coatings.

2 Introduction

building industry considerable use is made of extrusions that are anodized in standard lengths and subsequently cut to length, mitred, drilled, etc., for assembly into windows, double-glazing systems, shop fronts, etc.

This book includes guidance on the choice of material, design, surface pre-treatment, anodizing and colouring methods. Excellent and consistent service is obtained by careful attention to these choices.

The basic techniques for anodizing can be scaled down and simplified for demonstration in schools. On the industrial scale, however, the varied behaviour of different aluminium alloys and the control of processing demand considerable skill, knowledge and sound test procedures, especially as poor-quality anodic oxide coatings cannot always be detected by visual inspection.

Chapter 1

Principles of Anodizing

The manner in which anodic oxidation differs essentially from other industrial electrolytic processes will be apparent from the following three examples all using dilute sulphuric acid, say 10% by volume, as the electrolyte.

In Figure 1, if the electrodes are made of platinum or any other metal that does not dissolve at the anode or positive electrode, oxygen gas is liberated at the anode and hydrogen gas at the cathode. No metal is dissolved in the acid.

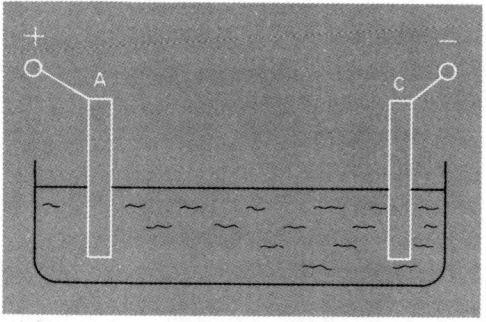

FIGURE 1. CURRENT ENTERING AND LEAVING SOLUTION IN ANODIZING

If the anode is made of copper it will dissolve in the acid and will be re-deposited on the cathode. This is the basis of many metal electrodeposition processes. In commercial production with soluble metal anodes little or no gas is evolved at the anode and cathode.

When the anode is aluminium, the cathode, in commercial practice, is either aluminium or lead. When current is passed the aluminium anode does not dissolve away like copper, nor is oxygen evolved in quantity. Instead, most of the oxygen that would have

4 Anodic Oxidation of Aluminium and Its Alloys

been liberated combines with the aluminium to form a layer of porous aluminium oxide. Hydrogen is liberated at the cathode.

The amount of aluminium oxide formed is directly proportional to the current density and time, i.e. to the quantity of electric current used. The progress of the formation of the anodic coating depends upon the chemical composition of the anodizing electrolyte and the chosen conditions of electrolysis. Some anodizing electrolytes have little or no solvent action on the oxide coating so that the process soon ceases, leaving a thin film usually referred to as a barrier-layer-type coating, the thickness of which is solely governed by the applied voltage and approximates to $\frac{1}{700}$ μm per volt. This type of coating is typically produced in solutions of borates, boric acid or tartrates.

If the electrolyte has some solvent action, then a porous film is formed and the oxidation process can continue leading to the production of relatively thick films, as for example in sulphuric acid. Eventually the rate of film formation is balanced by the rate of solvent attack, but this stage of the process is avoided in commercial practice.

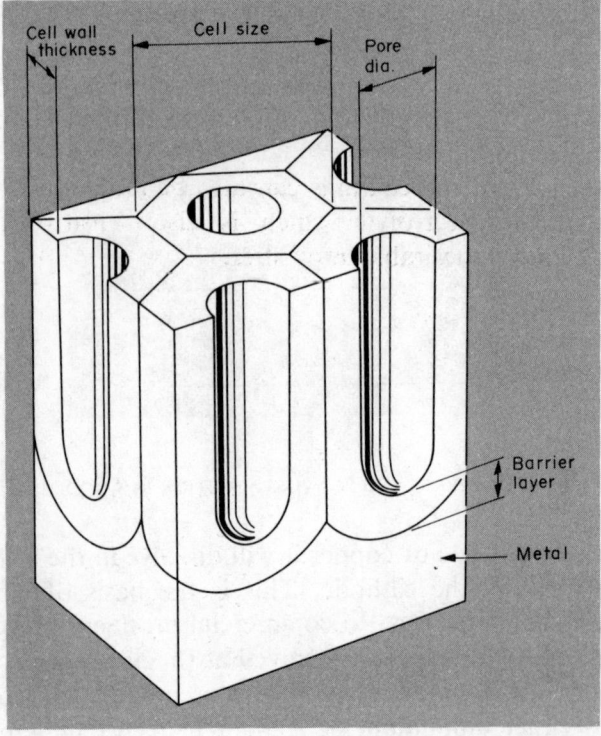

FIGURE 2. MICROSTRUCTURE OF ANODIC FILM

The structure of the porous type of anodic oxide coating is shown diagrammatically in Figure 2, and can be seen to comprise hexagonal columns each with a central pore which reaches down to a thin compact barrier layer which is continuously formed and transformed into the porous form during the process.

The diameter of the pores and the thickness of the barrier layer for any given electrolyte and temperature are proportional to the applied voltage. Thus by varying the anodizing conditions it is possible to alter the physical properties of the coating, such as the hardness, abrasion resistance and the density.

From the foregoing it will be appreciated that anodizing is a conversion process so that the appearance and other properties are completely dependent upon the composition of the aluminium and its surface condition. Anodizing, therefore, differs fundamentally from processes such as electroplating where a layer of metal is applied over the basis metal surface.

After anodizing the film is usually sealed by a hydration process to minimize the initial porosity. For some special applications physical sealing with an organic material such as oil is desirable. Colouring by immersion in dye solutions is carried out after anodizing but before sealing.

With aluminium alloys the alloying constituents are differently affected by the anodizing process, and in turn influence the appearance and structure of the coating.

The properties of the coatings can also be modified by chemical additions to the electrolyte which is also sensitive to certain adventitious and undesirable impurities.

Chapter 2

Applications of Anodized Aluminium

The properties of anodic oxide coatings on aluminium are unique among metal finishes and it is not surprising therefore that when commercial development blossomed in 1930 onward, anodized aluminium was hailed as a possible replacement for a wide range of materials. In due time many of these hopes proved unfounded but others received permanent acceptance in industry. New uses have been developed since that time, founded on new or improved processes described in this book.

Some of the more important uses are described in this chapter, roughly in their chronological order of promotion.

Anodizing as an undercoat for organic coatings

This was the first large-scale application and was based on the invention of the chromic acid (Bengough-Stewart) anodizing process. It was adopted as a standard finish for aluminium aircraft components and is still specified in the DEF 151 specification. This combination of an organic finish with an anodic coating ensures maximum life for the paint coupled with an underlying protective coating to provide further protection in the event of any paint failure.

Excellent service, even during sea-water immersion, is provided by sulphuric acid coatings, preferably sealed in a dichromate solution and then coated with an appropriate grade of paint.

The use of very thin coatings as a base for the subsequent painting and lacquering of continuous strip is also important.

Corrosion-resistant coatings

It was soon realized that unpainted chromic acid coatings,

Applications of Anodized Aluminium

especially when physically sealed with lanolin, had a high resistance to salt-spray corrosion and the use of this combination was also approved for aircraft components where painting would have interfered with the operational use of the coating.

By 1929 the sulphuric acid processes made their appearance with the ability to produce thick hard coatings which could be sealed with lanolin, oil, etc. The first Air Ministry approval of such a coating in the U.K. was granted in 1936.

The sulphuric acid coatings have been widely accepted for the treatment of aluminium exposed to marine and industrial atmospheres, and work thus processed before World War II is still in acceptable condition.

Today, the use of anodized aluminium for external and indoor building components provides the principal tonnage output for the world's anodizing capacity. This widespread application has depended on the invention of steam sealing and hot-water sealing without which the difficulties of physical sealing may well have inhibited such progress.

Typical examples of this class of work are: windows, patio doors, curtain walling, partition systems, double-glazing frames, canopies, grilles, fascias and shop fronts.

The corrosion-resisting properties of anodized aluminium have, not unexpectedly, found a wide field of application outside the aircraft and building industries. Many defence items that must survive long storage periods and others that need general protection in a tropical or marine environment are protected by anodizing. These include shell cases, tank armour, rocket components and fuses.

Coloured anodizing

Although the ability of anodic coatings to absorb dyestuffs was discovered in 1923 the commercial possibilities were not really exploited until the sulphuric acid process appeared in 1929. A wide range of dyestuffs was then selected and "coloured aluminium" as it was called posed a formidable threat to other finished metals and plastics. Some of the applications proved disastrous due to the poor light fastness of most of the available dyes but, on the whole, coloured anodizing secured a permanent and ever-increasing market. Early applications included ash trays, decorative metalware such as fruit bowls, tea trollies and panelling for indoor use. Larger-scale items included escalator metalwork and column casings in banks and stores.

The production of coloured designs by multicolour techniques attracted the attention of the label and nameplate industries who still offer this finish today.

Apart from the decorative use of colouring, it has also been applied to identify components or different alloys. For example, rivets or rivet wire have been anodized and dyed to avoid an incorrect selection of alloy rivets. Textile bobbins have also been dyed for yarn identification.

The problem of the poor light fastness of organic dyes began to be resolved in about 1937 when metal complex dyes of the "Neolan" class became available and an improved black dye Nigrosine D was developed. Some dyes offered by ICI Dyestuffs Ltd. included Solway Blue BS and Solway Ultra Blue, both of which were successfully used for shop fronts.

The use of pigment colours was well established in the early 1930s, but only the cobalt-permanganate bronze colour survived and is still used in some countries. This colour is used for outdoor metalwork.

In 1939 the brassy-gold colour produced by using ferric ammonium oxalate was developed in Germany but was little used until after World War II, since when it has appeared on many prestige buildings and has provided an economical substitute for gold-leaf-covered letters for shop fascias.

Integral colour bronzes appeared in 1959 and account for about 10% of the anodized aluminium used for buildings. Their hardness has also improved the performance of such items as door handles, fingerplates, door plates, shop fronts and entrances.

Electrolytic colours can be produced in many shades by using different colouring electrolytes. The majority of this class of work is produced in nickel, cobalt or tin solutions, all of which provide a range of light bronze to black shades having excellent light fastness. Here again, the building industry is the largest outlet.

Anti-marking applications

When aluminium alloys, unanodized, were introduced to the metalwork market for such applications as balustrades and grab rails the public complained of the greyish marking of light-coloured clothing or gloves that came into contact with the metal. Anodizing provided a complete answer to this problem and was adopted for example for all the metalwork in buses and coaches. In later years some of these items have been replaced by stainless steel. The successful promotion of aluminium knitting needles has been entirely due to the use of anodizing which prevents marking of the wool or the knitters' fingers.

Lighting equipment

Processes for chemical or electrobrightening aluminium did not appear until 1934. However, sulphuric acid anodizing was used to protect the surfaces of floodlights. The initial loss of total reflectivity was acceptable because this lower figure was maintained for years of service whereas plain aluminium continuously corroded and declined in performance. The anodized surface was easier to clean than corroded plain aluminium.

After the invention of the "Brytal" and "Alzak" electrobrightening processes the use of anodized aluminium especially based on high-purity sheet became an important material for the lighting industry. The advent of chemical brightening has facilitated the production of reflecting coatings on lower-quality sheet which is used for street lighting reflectors, illuminated ceiling grids and such specialized items as reflectors for operating theatre lights, aircraft lamps, airfield flare paths and miners' lamps. In World War II bright anodized flat sheets were used as shock-proof driving mirrors in tanks and many searchlight reflectors were similarly finished.

Heat reflection and radiation

Anodizing has long been used for aluminium electric-fire reflectors. The "Dimplex" unit with its heated silica tube is found in many homes. The easily cleaned surface withstands the humidity of bathrooms and is an efficient reflector of radiant heat provided that the anodic coating is restricted to about 1 μm.

In recent years the heat-radiating properties of thick coatings have been applied in the manufacture of heat sinks for electronic equipment. The coatings are often dyed black to increase their heat emissivity.

In addition to the natural colour reflectors some use has been made of coatings dyed in pale colours, particularly copper which enhances the appearance.

Reflectors for infra-red stoving ovens for paint have also been made in bright anodized aluminium where the reflectivity is only surpassed by gold.

Wear resistance and lubrication

Before hydration sealing became available the physical sealing with oils, waxes, etc., led to the rise of sealing with lubricating oils of coatings used for engineering purposes where an oil-wettable surface was an advantage. The principal application in this field was on anodized aluminium pistons for petrol and diesel engines. Sealing with graphite suspensions was also adopted. It was claimed that the coating had improved "running-in" properties and that the wear of the ring grooves in service was reduced. This type of sealed coating is still used on some diesel pistons and on air compressors, using "hard anodizing" to form the coating.

Hard anodic coatings are being very successfully exploited by the engineering industry for surfaces that have to withstand lightly loaded rubbing contact and where good corrosion resistance is necessary. Examples are cigarette manufacturing machines, hydraulic cylinders, coin-operated machine slides, textile spinning guides and ciné-camera components.

The coefficient of friction of anodized aluminium is appreciably lowered to about 0.1 by sealing with a suitable PTFE-resin mixture. This chemically inert sealant also provides additional corrosion resistance.

An interesting application of hard anodizing is for the treatment of the rollers that carry film stock during its manufacture. The wear and corrosion resistance are satisfactory and the complete anodized roller is cheaper than the stainless-steel rollers originally specified for this work.

As a general comment, the anodic oxide coating is much harder than the basis aluminium metal so that the general resistance to wear and marking is greatly increased by anodizing.

Electrical insulation

Although the anodic oxide coating is a good electrical insulator the danger of local breakdown due to minor defects has militated against any extensive use of anodized wire. However, continuously anodized aluminium strip has for many years been adopted for winding electromagnets and some types of transformers where weight saving is important. The oxide coating is much more resistant to heat than organic insulating media and is therefore selected for high-temperature environments.

Chapter 3

Factors Influencing the Choice of Grade of Aluminium for Anodizing

When selecting material for work that is eventually to be anodized the manufacturer's choice will be guided by the type of finish that is required. In some cases the mechanical properties of the aluminium will be of prime importance, e.g. aircraft components subject to stress in service. Here the choice will be limited and the appearance, attractive or otherwise, may have to be accepted.

Where resistance to atmospheric corrosion is the prime requisite it is often possible to select alloys giving coatings that combine good protection and an attractive appearance.

Many applications of anodizing are mainly decorative and here the choice of alloys is the most extensive, although for the production of very bright films, comparable with bright chromium plating, alloys of special high purity must be used.

It is evident therefore that any proposed scheme for the use of anodized aluminium can, with advantage, be discussed at an early stage with the anodizer who will eventually be entrusted with the processing.

Apart from the differences in the appearance of anodic film on different alloys, other variations can be introduced by mechanical or chemical pretreatment of the metal before anodizing. Indeed it is not unusual deliberately to modify the metal surface so as to obtain the best possible matching between different alloys, or between, for example, sheet metal, extrusions and castings.

In addition to the effects of varying alloy composition it is necessary to take note of any heat treatment that may be required to ensure the correct mechanical properties of the metal. Variations in heat treatment can modify the appearance of the coating.

12 Anodic Oxidation of Aluminium and Its Alloys

As a basic guide reference should be made to Tables 1, 2 and 3.

TABLE 1 ANODIZING CHARACTERISTICS OF WROUGHT ALUMINIUM

Material designation and temper	Suitability for anodizing			
	Protective	Colour	Bright	Hard*
1080A-0	E	E	V-E	E
-H8	E	E	V-E	E
1050 -0	E	E	V	E
-H8	E	E	V	E
1200 -0	V	V	G	E
-H4	V	V	G	E
-H8	V	V	G	E
2011 -TD	F	F (D)	U	G
-TF	F	F (D)	U	G
2014A-TB	F	F (D)	U	G
-TF	F	F (D)	U	G
2024 -TB	F	F (D)	U	G
-TD	F	F (D)	U	G
2618A-TF	F	F	U	F
3103 -0	G	G	P-F	G
-H4	G	G	P-F	G
-H8	G	G	P-F	G
3105 -0	G	G	P-F	G
-H4	G	G	P-F	G
-H8	G	G	P-F	G
5005 -0	E	E	E	E
-H4	E	E	E	E
-H8	E	E	E	E
5083 -0	V	V	G	E
-H2	V	V	G	E
-H4	V	V	G	E
5154A-0	V	V	G	E
-H2	V	V	G	E
-H4	V	V	G	E
5251 -0	V	V	G-V	E
-H3	V	V	G-V	E
-H6	V	V	G-V	E
5454 -0	V	V	G	E
-H2	V	V	G	E
-H4	V	V	G	E
6061 -TB	G	G	F	V
-TF	G	G	F	V
6063 -TB	V	V	G-V	E
-TF	V	V	G-V	E
6082 -TB	G	G	F	G-V
-TF	G	G	F	G-V
7020 -TB	F	F	†	G
-TF	F	F	†	G
7075 -TF	F	F	†	F

E = Excellent, V = Very good, G = Good, F = Fair, P = Poor,
D = Dark colours only, U = Unsuitable.
*Compared on the basis of a 50 μm film thickness.
†Variable response, depending on actual composition and heat treatment.

Factors Influencing the Choice of Aluminium for Anodizing

TABLE 2 ANODIZING CHARACTERISTICS OF CAST ALUMINIUM

Cast material designation		Suitability for anodizing		
BS	Alloy type	Protective	Colour	Bright
LM 0	Al 99.5	E	E	E
LM 2	Al Si 10 Cu 2 Fe	F	U	U
LM 4	Al Si 5 Cu 3	G	F (D)	U
LM 5	Al Mg 5	E	E	G
LM 6	Al Si 12	F	U	U
LM 9	Al Si 12 Mg	F	U	U
LM 10	Al Mg 10	E	F	U
LM 12	Al Cu 10 Si 2 Mg	F	F	U
LM 13	Al Si 11 Mg Cu	F	U	U
LM 16	Al Si 5 Cu 1 Mg	G	F (D)	U
LM 18	Al Si 5	G	F (D)	U
LM 20	Al Si 12 Cu Fe	G	F (D)	U
LM 21	Al Si 6 Cu 4 Zn	F	U	U
LM 22	Al Si 5 Cu 3 Mn	G	F	U
LM 24	Al Si 8 Cu 3 Fe	F	F (D)	U
LM 25	Al Si 7 Mg	G	F (D)	U
LM 26	Al Si 9 Cu 3 Mg	F	U	U
LM 27	Al Si 7 Cu 2	G	F (D)	U
LM 28	Al Si 19 Cu Mg Ni	U	U	U
LM 29	Al Si 23 Cu Mg Ni	U	U	U
LM 30	Al Si 17 Cu 4 Mg	U	U	U

E = Excellent, G = Good, F = Fair, P = Poor, U = Unsuitable, D = Dark colours only.

TABLE 3 SPECIAL BRIGHT TRIM MATERIALS

	Si	Fe	Cu	Mn	Mg	Cr	Zn	Ti	Others Each	Total
1080A	0.15	0.15	0.03	0.02	0.02	—	0.06	0.02	0.02	—
5005	0.30	0.7	0.20	0.20	0.5-1.1	0.10	0.25	—	0.05	0.15
5252	0.08	0.10	0.10	0.10	2.2-2.8	—	0.05	—	0.03	0.10
5657	0.08	0.10	0.10	0.03	0.6-1.0	—	0.05	—	0.02	0.05
6463	0.2-0.6	0.15	0.20	0.05	0.45-0.9	—	0.05	—	0.05	0.15

Sheet aluminium

The production of sheet by rolling results in a longitudinal grain pattern which varies with the alloy, and the degree of thickness reduction during rolling. It is technically difficult to produce sheets which when anodized will have a uniform finish free from objectionable defects such as rolled-in inclusions or an unattractive grain structure. However, a limited number of aluminium sheet producers can offer "anodizing quality" sheet which can be polished or etched and then anodized to give an acceptable finish repeatable from batch to batch. Strong alloy sheets are sometimes produced with a cladding of relatively pure aluminium which gives a good finish, but such clad sheeting needs careful treatment in order to avoid penetrating the layer of cladding during a polishing or etching process.

The materials in Table 3 are sheet alloys based on 99.7% aluminium plus the alloying material given in the Table. This range was especially developed so that after mechanical polishing and chemical brightening it is possible to anodize to produce very bright finishes, similar to chromium plating, that are widely used in the motor-car industry and for decorative components in the consumer durable market.

The alloys normally produced for bright anodizing are 1080A, 5657 and 5252. 1080A is used for small reflectors, where strength is not important, and for applications where severe forming is required. 5657, which contains about 0.8% magnesium in high-purity aluminium, is made in a range of tempers, and is suitable for most bright trim applications. 5252 has the same purity base but contains about 2.4% magnesium, to give it considerably greater strength, especially in the harder tempers. It is particularly useful for "snap-on" trim on vehicles.

Alloy 5005 is also made in anodizing quality, and is particularly suitable for satin and semi-bright finishes. In architectural work it provides an excellent match with anodized 6063 extrusions.

For extrusion, alloy 6463 is a high-purity base variant of 6063, especially developed for chemical brightening and anodizing.

The commercial development of anodizing processes that produce coloured anodic coating (the so-called "integral colour" processes) has resulted in the introduction of special alloy sheeting; the alloying additions facilitate the production of acceptable bronze and black finishes with a reasonable guarantee of uniformity from batch to batch.

Extrusions

Anodized extrusions form the biggest outlet for the anodizing industry. The favourite alloy is 6063 which has acceptable mechanical properties and can be anodized by all the conventional processes.

In order to minimize the number of pieces to be handled in the anodizing plant many of the extrusions are processed in standard lengths and afterwards parted and machined. The machined areas have no protective anodic coating, and this must be taken into account.

It is recommended that suppliers of extrusions should be advised that they are intended to be anodized. Special care is needed in the manufacture and maintenance of the extrusion dies in order to avoid prominent die lines that give an unattractive final finish. Die lines can be removed by grinding or polishing, but the cost of these operations may be prohibitive.

Forgings

Aluminium forgings are usually produced in the strong alloys to meet engineering specifications so that the alloy cannot be modified or substituted by another to suit a particular anodizing requirement. Where anodizing is used solely for crack detection the chromic acid process is used.

Castings

Apart from the general guide given in Table 2 account must be taken of the type of casting process employed, i.e.
   sand cast,
   permanent mould cast,
   die cast.

Castings that are to be anodized — particularly those to be colour anodized — must have a sound surface. Inattention to this point has led to many disappointments; if there is surface porosity, spotting and other imperfections in colour anodized films will occur due to the exudation of trapped electrolyte. The founder must therefore be made aware at the earliest stage that the required casting is to be decoratively anodized, so that he can give special attention to the surface quality. It is also possible to impregnate unsound castings prior to anodizing.

The choice of casting process depends largely on the quantity involved. In turn, the chosen process limits the range of alloys that can be used. The effectiveness of anodizing can be influenced by both the casting process and the alloy composition.

Sand castings, usually produced in short runs, have comparatively rough as-cast surfaces requiring considerable pretreatment if a smooth anodized finish is required. Permanent mould castings are smoother, but the smoothest surface is produced on die-castings, which, however, show surface flow-lines after anodizing. Irrespective of process, the area originally occupied by gates and risers may be visible after anodizing, even with careful preliminary polishing.

Practically all casting alloys can be anodized for protective purposes, but the dyeing and brightening processes are restricted to a smaller range of alloys. The common die-casting alloys are rich in silicon, so that their surface becomes grey after anodizing.

Where castings are to be bright anodized, homogenization or other heat treatment is useful.

Chapter 4

Factors Influencing the Choice of Anodizing Process

The properties of anodized aluminium depend on a combination of the following factors:
(a) The aluminium alloy.
(b) Pre-treatment processes.
(c) The anodizing process.
(d) Post-anodizing processes.

A very wide range of chemical solutions have been used or proposed as anodizing electrolytes. Most electrolytes are acidic, but some alkaline solutions have been investigated. The greatest tonnage of anodizing is produced in sulphuric acid-based solutions but other acids are used on a commercial scale in order to obtain special types of coatings.

Sulphuric acid

On all alloys except those containing insoluble constituents anodizing in sulphuric acid produces semi-transparent colourless films in thicknesses up to about 35 μm. When properly processed and sealed these coatings are suitable for both decorative and corrosion-resistant applications. The appearance of the coatings is considerably influenced by the original surface finish of the aluminium.

The so-called "bright anodizing" is achieved by selecting alloys based on high-purity (99.7% +) aluminium or alloys based on it, followed by a brightening process and then anodizing in sulphuric acid.

At low temperatures, e.g. −5°C to 5°C, the sulphuric acid process gives very hard coatings (known as "hard anodizing") and are widely employed in the engineering industry.

The sulphuric acid electrolyte can also be modified to produce films that will withstand a certain amount of forming without disrupting the film. This type of process is used for the anodizing (and colouring if required) of continuous strip.

Oxalic acid

Solutions of this acid tend to produce translucent hard yellowish films that have been used for architectural applications and also, in Japan, for cookware. The abrasion resistance is almost double that of a conventional sulphuric acid-process coating. This process is one of the earliest "integral colour processes" to achieve universal recognition.

Chromic acid

This acid was used in the first commercial anodizing process invented in 1923 by Bengough and Stuart. It produces thin films that are usually opaque grey in colour. It is still widely specified for the treatment of aircraft components for four reasons in particular:
1. It is a good base for subsequent painting.
2. The minimum amount of aluminium is converted during the process, thus reducing any loss of metal thickness on thin stressed sheet components.
3. The loss in fatigue resistance is lower than in the case of sulphuric acid.
4. If chromic acid becomes trapped in, for example, riveted or overlapping joints it is less likely than sulphuric acid to cause corrosion.

Chromic acid anodizing is also used for crack detection, e.g. in forgings. The orange/red electrolyte oozes from cracks after anodizing and colours the dry coating.

Phosphoric acid

The anodic films produced in phosphoric acid have larger pore diameters than in conventional sulphuric films. This greater diameter provides a better conducting path and it is for this reason that phosphoric acid films have been used as one of the methods for pretreating aluminium prior to electroplating. During the latter process the anodized aluminium is used as the cathode and plated, for example, with nickel. The phosphoric acid process is rarely used as a finish in its own right due to the existence of more convenient processes.

A closely defined phosphoric acid-anodizing process has recently been adopted to prepare aluminium for adhesive bonding for aircraft components.

Integral colour anodizing electrolytes

This term covers a wide range of organic acid solutions, usually containing small additions of sulphuric acid. The films produced are over twice as abrasion resistant as sulphuric acid films and can be produced in a range of colours from pale gold/beige through bronze to black.

These processes have been widely used for architectural work such as windows, shop fronts and curtain walling. The hardness of the film has also resulted in its use for other industrial purposes as an alternative to "hard anodizing" and has recommended itself for building components exposed to the "sandy" atmosphere of the Middle East.

The precise colour produced by a given integral colour process depends upon the alloy composition. It is advisable to specify a single source of supply for extruded sections, for example, or if this is not feasible, to ensure that supplies from different sources are segregated when sent for anodizing.

The integral colours are exceptionally light fast and can confidently be specified for prolonged exposure outdoors.

Non-solvent electrolytes

The acid solutions mentioned above have a solvent effect on the oxide film and the resultant porosity enables relatively thick films to be built up. The non-solvent electrolytes, e.g. boric acid or ammonium tartrate, provide thin impervious barrier layer films with electrical properties suitable for use in capacitors. This type of process is generally carried out on continuous strip and has a restricted but very important field of application.

CHOICE OF PROPERTIES OF THE ANODIC OXIDE COATINGS

To assist in the choice of suitable coatings their properties are briefly described in this section, and then discussed in greater detail in Chapter 14.

Appearance (Texture)

This will largely depend upon the state of the aluminium surface before anodizing. The subject of the mechanical finishing of aluminium surfaces is very fully dealt with in a companion volume. Anodizing itself causes some etching of the aluminium and will reveal the grain structure of the metal.

As a general guide, smooth aluminium surfaces produce smooth anodic coatings, whilst surfaces roughened by etching, grinding, scratch brushing and abrasive blasting all give rise to rough coatings. These considerations apply to all types of alloys and all anodizing processes regardless of the electrolyte used.

Reflectivity

The optical properties of anodic coatings are influenced not only by the alloy used but also by the anodizing process. Where high total or specular reflection values are needed it is obvious that coatings of maximum transparency shall be used, and for this reason one of the sulphuric acid electrolytes is chosen. The effect of alloy composition on reflectivity is discussed in Chapter 14.

Colour

Coloured coatings are produced by
1. dyeing in organic dyestuffs,
2. impregnating with coloured pigments,
3. colouring by electrolytic means,
4. using "integral colour" anodizing processes.

As in the case of reflectivity the first three of these methods give the best results on the almost colourless sulphuric acid coatings. Where the range of colours produced by the "integral colour" process is acceptable then the appropriate process can be selected.

Dimensional changes

The anodic coating occupies a greater volume than the metal from which it is formed so that in most cases there is a tendency for overall dimensions to increase. A growth figure of 25-50% of the final coating thickness is often achieved. For example a 25 μm film coating will give a net growth of about 8 μm. These increases are maximized by using anodizing electrolytes that give porous films but which have a relatively low solvent action on the coating, i.e. "Hard" anodizing processes.

Dimensional change can be minimized by specifying the chromic acid process or by reducing the film thickness in one of the other electrolytes. The choice of electrolyte will then depend upon the appearance and other physical attributes that are needed.

Corrosion resistance

The development of weather-resistant coatings calls for processes that will produce thick coatings of good quality and correctly sealed. The sulphuric acid and "integral" colour processes therefore predominate this field of application. Chromic acid anodizing, particularly when painted, is specified as a corrosion-resistant coating for aluminium aircraft components.

Abrasion resistance

The conventional sulphuric acid process gives coatings that have fair resistance to abrasion, but this valuable property can be enhanced by using "hard" anodizing processes based either on cold (−5 to +5°C) sulphuric acid or by using oxalic acid or one of the "integral" colouring electrolytes. The chromic acid coating, although inherently hard, is too thin to provide good abrasive wear resistance.

Heat radiation and absorption

The emissivity of most metals is usually less than 10% of a perfect black body. The formation of an anodic coating, irrespective of the anodizing process, causes a rapid increase of emissivity which rises to a figure in the region of 80% at a coating thickness of 2.5 μm or more. In practice, the sulphuric acid and "integral" colour process are both used.

Heat reflection

The infra-red reflectivity of anodized aluminium falls sharply with an increase of coating thickness and it is customary to restrict coatings to a maximum of 0.8 μm in order to secure a reflectivity exceeding 80%. If a highly specular infra-red reflecting surface is needed then "bright" anodizing is carried out on suitable high-purity metal that has first been mechanically polished. The sulphuric acid process is usually specified for this class of work.

Electrical insulation

Although the porous type of anodic coating conducts current continuously during anodizing, the dry coating, sealed either by a hydration process or with an organic filling, is a useful electrical insulator. Thick coatings produced in sulphuric acid under special conditions give acceptable insulation. Oxalic acid anodizing also contributes to this class of coating. Special proprietary electrolytes have also been developed for the continuous anodizing of aluminium wire for use in electrical windings.

Chapter 5

Designing for Anodizing

This chapter deals with the effect of mechanical pre-treatment processes and fabrication practice on the design of the final anodized product, and points out some of the undesirable features that should be avoided. It cannot be too strongly emphasized that the anodizer should be consulted at the earliest stages of design in order to avoid errors that will lead to an unsatisfactory product. The pre-treatment processes are described in detail in a companion volume.

Mechanical pre-treatment processes

Mechanical polishing will produce equally smooth surfaces on all forms of aluminium provided that the surfaces are accessible to the polishing mops available. It does not follow that the same degree of smoothness will remain after anodizing. Sheet and extrusions lose little gloss when anodized, but castings are liable to exhibit a pronounced and sometimes unsightly grain structure, whilst forgings reveal the flow pattern of the metal.

For the production of polished decorative finishes reference should be made to Tables 1, 2 and 3. In order to provide a matching finish between sheet, extrusions and castings it may be necessary to specify different coating thicknesses on the various materials.

When selecting polishing compositions for aluminium one must avoid those containing green chromium oxide or red rouge. These coloured compounds become embedded in the metal and cause a green or red stain in the anodic coating.

Sand and vapour blasting

The effect of both these processes is to disturb the metal surface to give a degree of matting, varying with the size of grit being used. The more attractive surface is produced by vapour blasting using a jet of wet abrasive slurry. Using anodic coatings in the 5-10 μm range vapour blasting provides a useful method for obtaining a reasonable match on all forms of aluminium (except pressure die castings) that give "colourless" coatings.

The blasting processes are carried out inside a chamber, the size of which limits the dimensions of the article being treated. The designer must duly note this limitation.

Grinding, banding and brushing

By the use of mops dressed with abrasives, or endless abrasive coated bands or fibre brushing wheels a directional lined texture is produced on aluminium. The banding method is suitable for flat surfaces whilst the relative flexibility of mops and brushes indicates their use for more irregular surfaces. All three of these methods are useful for obtaining a matching appearance between wrought and cast metal.

A special application of grinding is the finishing of aluminium spinnings and other items having a symmetrical longitudinal axis by holding garnet or "Aloxite" paper with paraffin lubricant against the metal whilst revolving on a lathe. The lined finish covers metal defects and disguises the distorted grain structure of spun and pressed sheet metals.

Scratch brushing

Satin finishes can be produced by holding aluminium against revolving stainless-steel wire wheels. Mild chemical etching is then required to clean the surface prior to anodizing. Generally speaking, the scratch brushing process is used on sheet metal in a continuous coil on specially designed automatic machines. This particular product can be anodized without pre-etching.

FABRICATION PRACTICES

Drawing, forming and spinning sheet metal

Sheet aluminium has a directional grain running along the direction of rolling. Even after polishing either by bright rolling or by mechanical polishing, the grain will reappear during chemical pretreatment or anodizing. This effect can be tolerated on flat or formed sheet provided that the designer specifies that the sheet is cut with the grain in the best direction to suit the appearance of the finished product.

Designing for Anodizing

In the case of drawing and spinning the grain must be distorted and may therefore become unsightly. This defect can be minimized by using "anodizing quality" sheet or "bright trim" sheet. The latter grade of metal, as its name implies, is specified for "bright anodizing".

Bending of large sections

Some classes of architectural metal work, such as handrails, call for the forming of large extruded sections where it is necessary to heat the metal to facilitate the bending operation. On heat-treatable alloys this heating may cause the precipitation of alloy compounds that do not dissolve away clearly during anodizing, thus giving a greyish or brownish discoloration. This defect can be dealt with by re-heat treating the bent section or alternatively the trouble can be avoided by specifying a non-heat-treatable alloy.

WELDING PROCESSES
Gas, arc and flash-butt welding

As a general rule the filler rod used for gas welding must be selected so that its appearance after anodizing matches as closely as possible that of the parent metal (see Table 4). Nevertheless, the weld puddle, having a cast structure, will be more or less visible when anodized. On heavy sections, where the design permits, the welding can be carried out at the back only, leaving a hair-line joint at the front.

When certain aluminium-magnesium-silicon wrought alloys are welded, the heat causes re-crystallization and structural changes in the metal near the weld, resulting in the formation of a brown surface band on anodizing (Figure 3). This is usually ½-1 in. away from the weld, depending on the amount of heat flow along the section, and is additional to any difference of appearance in the weld. The stained area depends largely on the heat input, but in hand processes the skill of the welder has some effect. Gas welding may give extensive effects; inert-gas arc welding is less harmful; flash butt welding gives the best results. Of the alloys in the group the effect is least on 6063 and 6082, but variations in the intensity of the effect may be found even within material to the same specification. Stripping and re-anodizing may reduce the intensity of the mark, but complete re-heat

TABLE 4 FILLER AND ELECTRODE WIRES
For optimum weld properties and colour match after anodizing

Parent metal	Recommended wire
Wrought products BS 1470	BS 1475
1080A	1080A
1050A	1050A
1200	1050A
3103	3103
5251	5056A
5154A	5154A
6063	
6061	5056A
6082	
Cast products BS 1490	
LM 5	5056A
LM 6	4043A

FIGURE 3. WELDED JOINT IN ANODIZED ALUMINIUM, SHOWING STAINING

treatment after welding and before anodizing is necessary to prevent its occurrence.

Individually anodized components can be subsequently fusion welded provided extreme care is taken, the design is a good one, and

Designing for Anodizing 27

the welding is done by one of the gas-shielded processes from the reverse side of substantial sections; fine-wire welding techniques are also useful in joining already anodized parts. However, the fusion welding of anodized articles should be attempted only when it would be quite impracticable to weld before anodizing. It should be realized that the anodic film crazes under the welding heat and the protection it affords is then much lower, although the crazing may not be readily visible after cooling.

Parts to be cold pressure welded receive a good surface if individually anodized beforehand. Break-up of the anodic film under the extensive deformation exposes clean aluminium surfaces which readily weld together.

Spot welding

The indentation caused by a spot weld can be smoothed out by polishing, but is best disguised by one of the mechanical "roughening" methods. There is also the danger of trapping anodizing electrolyte between the faces of the spot-welded joint. On the whole it is advisable to locate the spot welds in "non-significant" areas.

Riveting

A riveted joint will always remain visible. It has the same disadvantages as spot welding and is even more liable to exhibit stains due to the escape of trapped electrolyte from between the metal interface and from the rivet hole. The colouring of anodized riveted joints using organic dyes usually results in colourless areas around the rivets and joints. Where the grey colour of the coating is acceptable the chromic acid process is preferred for riveted components.

Adhesive bonding

Joints bonded with adhesive ("resin bonding") may be satisfactorily anodized provided the adhesive is not present on the surface and provided the joint is sound, so that capillary effects cannot occur. Separate electrical connections must be made to each component in the assembly, as the resin film is a good electrical insulator. However, it is much sounder practice to anodize the components preparatory to resin bonding, as the bond strength is generally similar on anodized and unanodized material.

OTHER DESIGN CONSIDERATIONS

The dangers that may arise from trapped electrolyte lead to a general recommendation to avoid designs that are liable to this defect. Examples are:

1. Blind holes, especially less than, say, 6 mm diameter.
2. Closed beaded edges on spinnings.
3. Very narrow (less than 2 mm width) channels in extruded sections.

Hollow items are liable to form airlocks inside giving rise to unanodized areas. Careful attention to the jigging of the work during anodizing usually cures this problem. However, there are occasions when, for example, welded window frames are made from hollow extruded sections. In this case 12 mm holes must be drilled in opposite corners of the frame to allow electrolyte to drain out. It is not sufficient to rely on the liquid tightness of the welds.

The technique of anodizing and colouring has advanced to the stage where acceptable limits for the matching of components can be agreed between the purchaser and the anodizer. However, perfect matching is not commercially feasible and it is therefore worth bearing in mind the possibility of using contrasting finishes for adjacent areas. For example, large areas of curtain wall sheet can either be broken up with other materials such as stone or concrete or by "break strips" in another colour or texture.

All articles to be anodized have to be suspended on jigs during treatment and must be provided with adequate electrical contact areas to carry the anodizing current. These areas should be arranged on non-significant surfaces as they have no protective anodic coating on them. If difficulty is anticipated in allowing for contact areas, the anodizer should be consulted at the design stage.

Chapter 6

Anodizing Equipment

The Anodizing Shop

Premises

If there is a choice of existing premises or if new premises can be built there are certain features that are desirable for housing an anodizing plant. It must be realized that anodizing involves the use of aggressive chemicals, corrosive and even toxic fumes may be evolved, and the drainage from the process needs special consideration.

The general construction should have brickwork or concrete walls, concrete roof beams and a double-skinned asbestos roof with glass fibre in the "sandwich". The brickwork should be painted in white (this aids inspection for colour matching).

The floor should be in reinforced concrete with a slope to an acid-proof gulley drain or drains, and covered with acid-proof tiles (and jointing) or with an acid-resisting resin coating. The gulley drains should be connected to a suitable effluent-treatment plant (see Chapter 16). In order to confine any liquid spillage to the protected floor area a bund wall, say 15 cm high, should be built around the perimeter.

North window lighting (in the northern hemisphere) will aid inspection by day, whilst by night the "colour matching" class of fluorescent-tube lamps is essential.

If an existing building with steel roof trusses, etc., has to be used all the steelwork should be cleaned and coated with a good anti-corrosive system — it is strongly recommended that a specialist firm should be employed for this work. Any subsequent corrosion of the steel should be regularly attended to by local cleaning and repainting.

Public Services for Anodizing

Electricity

Anodizing is a large consumer of electricity and an adequate supply (including projected future requirements) must be available. Power-factor-correction equipment is a profitable investment and should be investigated after the plant is fully operational. It may also be financially advantageous to take a main supply at high voltage (11,000) instead of the normal 400 V tension, especially for large installations of, say, ½ MW upwards. Most electricity-supply contracts are based on a maximum kVA demand rate, and in view of the load fluctuations imposed by a number of separate anodizing tanks a maximum kVA alarm system will help to avoid unnecessary and expensive overloads on the system.

Water

Apart from the water needed to make up the various chemical solutions, a good supply is required for rinsing work between each stage of the process. Therefore, a storage cistern with capacity for, say, 6 hours plant operation is needed and a mains intake pipe of suitable size to cope with the demand. The purity of the water is also of importance for certain of its uses in anodizing. A typical analysis should be obtained from the water-supply company and, in particular, any information on seasonal fluctuations in quality (see Chapter 11).

The pressure from the water cistern will usually be insufficient to operate top spray systems, in which case a rotary pump will be needed to provide, say, a 60 psi pressure on the water line supplying the sprays.

Drainage

It is obviously essential that the local drainage facilities should be capable of dealing with the anticipated output from the plant and that the connecting pipe to the main sewer should be adequately sized.

At an early stage an agreement will have to be made with the Water Authority for the discharge of effluent. This agreement will doubtless impose conditions regarding the volume, chemical composition, temperature, etc., of the effluent, and these requirements will in turn govern the design of any effluent-treatment equipment (see Chapter 16).

ANCILLARY SERVICES FOR ANODIZING

Process Heating

Steam

Most anodizing plants are heated by immersed steam heated coils supplied from a boiler which may be fuelled by oil, gas, solid fuel or electricity (electrode-type boiler). The choice of boiler will depend on the size of installation, the relative prices and convenience of the alternative fuels. A steam pressure of 100 psi at the boiler with 60 psi in the plant is usual. The condensate should be returned to the boiler but it is essential to provide a pH test system for the hot well so that warning can be given of any contamination of the condensate from leaking steam coils.

Gas and electricity

Some of the smaller plants do not warrant a boiler installation, in which case the tanks are designed for external heating by gas or internal heating using immersion heaters or cables. To minimize heat wastage with external gas heating the tanks can be provided with horizontal tubes inside the tank in which burning gas is injected at one end and taken to a flue at the other.

The covering or casing for electrical immersion heaters must be carefully chosen to withstand corrosion attack by the heated solution.

Process Heat Conservation

All forms of heating are becoming increasingly expensive so that expenditure on heat conservation is a good investment. The following items particularly merit attention:

1. All steam mains and piping should be adequately lagged.
2. All internally heated tanks should be lagged on the outside, including the bottom.
3. Where practicable, hot tanks not in use should be covered.
4. Hot tanks in constant use, especially at 90°C or over, should have mechanically operated covers or, if convenient, be covered with a double layer of polypropylene spheres.
5. Any reduction in the efficient operating temperatures of heated solutions will reduce heating costs. For example, cleaning solutions have been formulated to operate at lower than usual temperatures.

Anodic Oxidation of Aluminium and Its Alloys
Agitation of Solutions

It is essential to provide good agitation in the anodizing solutions that operate at a high wattage per unit of surface area. For example, a typical sulphuric acid process will demand 16 volts 1.5 amp/dm^2, i.e. 24 watts/dm^2, whilst an integral colour process may need as high as 100 volts (but more typically 70 volts) at 3 amp/dm^2, i.e. 300 watts/dm^2. The conventional chromic acid process demands 12 watts/dm^2 maximum and only needs sufficient agitation to prevent temperature "layering" in the bath.

Compressed air

Compressed air is the most popular and effective method for agitating anodizing solutions using perforated pipes at the bottom of the solution. A high-pressure supply is unnecessary and wasteful. It is only necessary to overcome the hydrostatic pressure of the solution and to supply an ample volume of air. A typical supply pressure is 10 psi. Rotary compressors are suitable for this work and should be designed to deliver oil-free air. If circumstances permit, the compressor should be installed above the top level of the solutions being agitated or alternatively a small hole (3 mm diameter) should be drilled at the point where the air pipe passes over the edge of the tank — thus avoiding the danger of a siphon "suck back" of solution into the air line after shutting down the compressor.

Mechanical agitation

Liquids of a viscous nature such as chemical polishing solutions and viscous caustic soda etching solutions, i.e. those having a high dissolved aluminium content, are difficult to agitate with air. "Blind areas" are liable to occur on upward-facing surfaces or in recesses. This problem is best overcome by using mechanical agitation equipment to provide a vertical reciprocating movement with a stroke of, say, 10 cm at a rate of 20-40 strokes per minute. The drive system for oscillating the work-carrying rod may be actuated pneumatically, hydraulically or purely mechanically, depending on the scale of operation.

This oscillating system is also very useful for electropolishing and for those electrobrightening processes that call for agitation, but in these cases a flexible system for supplying current to the moving work rod must also be incorporated.

Cooling systems

Reference has already been made to the wattage per square decimetre consumed during anodizing. The resultant heat evolved must be continuously removed from the anodizing solution in order to maintain a constant temperature. The two most popular cooling systems use either immersed coils or panels or, alternatively, the anodizing solution is pumped from the tank through a heat exchanger and then back into the tank through a sparge pipe along the bottom of the bath.

The cooling medium may be either low-temperature water from the mains or a natural water source or a liquid cooled by a refrigerating unit. The choice of cooling system is governed by the total heat load on the system and the availability or otherwise of a sufficient volume of low-temperature water, for example, at 12°C or lower throughout the year. The maximum rate of heat evolution can be calculated from the maximum kilowatt figure demanded by the process when operating at maximum amperage and maximum voltage. For example, a sulphuric acid anodizing bath working at 5000 amperes at 15 volts will consume 75 kW. 1 kW = 3.6 MJ/hour. In the United Kingdom it is customary to specify refrigeration requirements in "tons" which can be calculated from the formula:

$$\text{Tonnage of refrigeration} = kW \times 0.3$$

which also takes account of the heat of formation of aluminium oxide.

The heat-removing capacity of the cooling system for each tank must be designed to deal with the maximum load in that anodizing tank, but where a number of anodizing tanks have to be cooled there is a considerable diversity factor due to some tanks being off-load whilst others are on-load. No heat is evolved, for example, whilst work is being loaded into and removed from the tank. For a multiple-tank installation it is advantageous to have a common reservoir of cooled liquid from which supplies are pumped to each anodizing tank as and when demanded. The refrigeration capacity for cooling the reservoir may well be less than 75% of the total theoretical demand of all the anodizing tanks.

A well-designed cooling system with good thermostatic controls should be capable of maintaining a given anodizing solution temperature within ±1°C, always provided that the solution is kept well mixed by vigorous agitation.

If it is practicable to use the water supply for cooling it can afterwards be taken to the rinse tanks before being discharged to the drain.

Heat exchangers for cooling anodizing solutions were originally made in chemical lead, then in graphite and more recently in stainless steel. Exchangers designed in stainless steel are compact and efficient. In operation, the acid side of the exchanger is kept in continuous circulation, so avoiding the possibility of blockage by the freezing out of aluminium salts through reducing the temperature of stationary acid in the exchanger to too low a temperature. The thermostatic control is actuated by a probe in the anodizing solution and in turn operates a pump to supply cooled liquid through the "coolant" side of the exchanger.

Anodizing current supply

Most anodizing processes operate with direct current which is usually supplied by a rectifier. Modern rectifiers are constructed with silicon diodes and when operated from a three-phase supply are arranged to give a "hexaphase" output which minimizes ripple — a figure below 5% ripple on full load is achievable. Excessive ripple is the result of too large a residual AC component in the DC output and can give rise to the formation of faults in anodic coatings. Defective coatings can also be produced if one phase of the three-phase circuit fails.

Water-cooled rectifiers are available for use where the ambient temperature is high — for example, under tropical and semi-tropical conditions. In any case rectifiers should be installed in a clean well-ventilated area as close as possible to the anodizing tanks but separated from them by a partition or wall.

Some versions of hard anodizing processes use AC superimposed on DC. In this case the DC output of a rectifier is connected in series with the secondary winding of a single-phase transformer and the anodizing tank. The current-carrying capacity of the secondary transformer winding must be sufficient to carry the total AC and DC amperage loads.

In small manually operated plants the current supply to the anodizing tank is often operated by a manual control, but even in small installations, and certainly in large ones, some form of automatic control is desirable in order to standardize the processing conditions. A typical control unit when activated will increase the anodizing voltage from zero to produce a selected voltage or amperage figure and will automatically maintain this figure for a pre-selected time after which the voltage will be reduced to zero and signalled by light or bell. For the more complex anodizing processes using a succession of selected voltages and currents, fully programmable control units are available, thus dispensing with the constant attendance of the anodizing operator during the processing period.

Electrobrightening current supply

The first electrobrightening processes for aluminium—"Alzak" and "Brytal"—were invented in the middle 1930s at a time when the necessary DC current was provided from DC dynamos. At that time metal rectifiers were beginning to replace dynamos as they offered the advantages of greater electrical efficiency and lower maintenance costs. However, the rectifiers then available produced DC with a pronounced ripple which reduced the effectiveness of the brightening process. The quality of DC supplied from modern rectifiers operating from a three-phase supply is, however, suitable for electrobrightening. Rectifiers operating on a single-phase supply will not give satisfactory results, a point to be borne in mind when setting up very small-scale or laboratory units.

Electrocolouring current supply

A later chapter describes the technique of electrocolouring where a wide variety of current forms is claimed in the very extensive range of patents in this field. The earlier commercial processes, such as "Anolok", use a single-phase AC supply with a maximum output of 30 volts and an operating current density of about 0.3 amp/dm^2. An important feature of this equipment is a control panel to provide pre-selected voltages for given periods of time. Other versions of this type of process use special AC waveforms, a mixture of AC and DC or plain DC. In some cases the special equipment needed is supplied by the licensors of the process.

Fume extraction

In order to comply with the Health and Safety at Work Act, any unpleasant or noxious fumes created in the anodizing and ancillary process must be exhausted and, if necessary, suitably treated before discharge to the atmosphere.

The total amount of air exhausted from the plant, together with any air exhausted by way of general shop ventilation (e.g. by roof fans), needs replacement with fresh air from outdoors; in the winter months this replacement supply must be pre-heated to an acceptable ambient temperature of, say, 18°C.

Table 5 summarizes the exhaust requirements for various processing solutions.

TABLE 5

Process	Exhaust	Scrubbing required	Duct material
Alkaline cleaning	√	×	Steel, plastics
Hot acid cleaning	√	×	Plastics
Chemical polishing	√	√	Stainless steel, GRP, plastics
Caustic etching	√	√	Steel
Electrobrightening (alkaline)	√	√	Steel
Electropolishing (phosphoric/chromic acid type)	√	√	Stainless steel
Anodizing: sulphuric acid	√	Optional*	Plastics, Monel metal
chromic acid	√	Optional*	Steel
integral colour	×	×	*Plastics
Nickel acetate sealing	√	×	Plastics
Hot-water sealing	√	×	Plastics

*The spray arising at the cathode in these solutions can be considerably reduced by adding a spray-suppressing agent.

Tank construction

Some of the materials now commonly specified for tank construction are listed in Table 6.

The tanks themselves should be designed so as to avoid undue distortion when filled with liquid, for example in a 10-metre-long tank the total deflection at the centre should not exceed 5 cm. Tanks that require frequent cleaning out, e.g. water rinse tanks, should have a drain valve at the bottom, whilst tanks containing strongly alkaline or acid solutions are safer if no bottom valve is fitted. They can be emptied using hoses and a suitable portable pump.

The heating coils are usually fitted along the bottom of the tank, except when appreciable amounts of sludge are likely to accumulate. For example, caustic etching and chemical polishing tanks are usually designed with heating panels or coils fitted along the vertical walls.

The top flanges of tanks should be provided with a hardwood frame or a plastics coating in order to prevent direct metallic contact

TABLE 6 MATERIALS FOR TANK CONSTRUCTION

Solution	Tank material
Trichlorethylene	Galvanized iron (special construction required)
Hot sulphuric acid cleaner	Chemical lead-lined steel
Caustic soda	Mild steel
Etches containing HF	Plastics-lined mild steel or lead lining (in the absence of nitric acid)
"Brytal"	Mild steel. Separate mild steel cathodes
Phosphoric acid base electropolishing solutions	Antimonial lead-lined steel, preferably with glass shields on ends and bottom. Separate lead or stainless-steel cathodes
Phosphoric/sulphuric acid-based chemical polishing solutions	Chemical lead- (not antimonial) lined steel
Phosphoric acid/nitric-base chemical polishing solutions and rinses	Molybdenum-bearing stainless steel for tanks and steam heating coil, stainless or silica-sheathed immersion heaters
Sulphuric acid anodizing	Polythene or PVC linings, acid-resisting rubber- or lead-lined steel or acid-resisting bricks (lead or lead-covered steel coils). Lead or aluminium cathodes
Chromic acid anodizing	Mild steel with stainless-steel cathodes
Oxalic acid anodizing	Hard rubber or PVC-lined mild steel
Nitric acid (cold)	PVC- or butyl rubber-lined mild steel, stainless steel or glazed earthenware
Water rinses	Reinforced concrete; lead-, rubber-, plastics-lined mild steel, galvanized mild steel (not suitable for acid rinses), wood, acid-resistant brick, PVC or fibreglass
Aqueous sealing and fixing solutions	Stainless steel, chemical lead-lined steel
Nickel and cobalt acetate	Stainless steel, rubber- or lead-lined steel
Potassium permanganate	Stainless steel or aluminium
Ferric ammonium oxalate	Rubber-lined or lead-lined steel; stainless steel
Organic dyes	Stainless steel. Rubber-lined steel or wood may be used where bath is kept for one dye only
Chrome/phosphoric stripping or desmutting	Stainless steel

between the work bar and the tank. This contact would facilitate the flow of a small galvanic current which causes "spotting-out" in some dye solutions.

In both manually and hoist-operated plants the solutions dripping from work during transfer from tank to tank (known as "drag-out") should be caught on plastics capping fitted over the gaps between each tank. The capping should be sloped in the opposite direction to that of the work travel.

Tanks are best supported on brick, concrete or wood so as to leave a gap underneath of at least 15 cm. The protection against corrosion of the outside of tanks is an important item to be considered. Plain mild-steel tanks should be shot-blasted and then provided with a properly designed paint or plastics-coating system which should also be applied to any outside supporting steelwork. The life of such coating systems is greatly prolonged by using the capping devices mentioned above.

Water-purification plant

At an early stage in the design of an anodizing plant an analysis of the water supply should be obtained. Certain natural impurities in water can render it unsuitable for making up some of the processing solutions. For example, chloride can upset the operation of anodizing solutions, while excessive phosphate and silicate figures will prevent effective hot-water sealing. A high total dissolved solids content will cause water staining during the final drying off of work.

Three methods of water treatment are used: (1) Base exchange process. (2) Ion exchange process. (3) Reverse osmosis.

The base exchange process, commonly known as water softening, removes calcium and magnesium ions from the water and this is often sufficient treatment for water for dyeing and for boiler feed. The equipment is available with manual or automatic regeneration.

For the removal of virtually all impurities from water, a de-ionizing plant is required. This may have separate beds of anion and cation removal resins or mixed beds of both resin types. This equipment is expensive but very effective — it is available with automatic regeneration.

Reverse osmosis is a more recent development in ion removal. It is especially useful where there is a high level of impurities in the water — for instance in some bore-hole supplies. Reverse osmosis is sometimes recommended as a preliminary purification process before final de-ionizing. The effect of this two-stage arrangement is to reduce considerably the work load in the de-ionizing unit and to decrease correspondingly the frequency (and cost) of regeneration.

The following are guide figures for the maximum limits tolerable for various processing solutions:

For Sulphuric acid anodizing	Chloride as NaCl	20 mg/l
Chromic acid anodizing	Chloride as NaCl	20 mg/l
	Sulphate as H_2SO_4	50 mg/l
Integral colour anodizing	Chloride as NaCl	10 mg/l
	Fluoride as NaF	5 mg/l
Hot-water sealing	Silica as SiO_2	10 mg/l
	Phosphate as H_3PO_4	5 mg/l

Other impurities can be detrimental to hot-water sealing and it is good practice to carry out sealing tests on samples sealed in the proposed water supply (see Chapter 13).

Water treatment is an important subject, and if there is any doubt about the suitability of a given supply reference should be made to water-treatment specialists.

Plant Types and Layout

The early anodizing plants were manually operated. With the advent of automatic electroplating plants in the 1930s there was some movement towards adapting the same design for anodizing, but the real increase in automatic anodizing plants occurred after World War II. In the United Kingdom there are over 400 anodizing installations with amperage capacities varying from about 500 to 80,000, but it is in the intermediate size of plant that full automation is mostly applied. Some of the first examples in this range were the Hoover Ltd. plant for anodizing washing-machine bodies and the Frigidaire Ltd. plant for anodizing ice-cube trays and other refrigerator parts. Later installations dealt with bright trims for motor cars and other consumer durables, and included a computer-controlled unit for aluminium extrusions.

For short runs of work in a variety of finishes a manual plant is commonly used, but when there are longer runs, a small variety of finishes and, in particular, where the work load is physically heavy, an automatic or a manually controlled hoist-operated installation is indicated. A major demand for anodizing arises in the architectural and home improvements market where much of the metal is processed in the form of standard lengths of extrusion. This work is mainly dealt with in manually or console-controlled hoist-operated plants.

40 Anodic Oxidation of Aluminium and Its Alloys

A very special type of automatic unit is that for the processing of continuous lengths of aluminium strip or wire (see Chapter 9).

The style of layout is too often governed by the shape of the space available, which may not be ideal for the purpose. It is good practice to draw the original layout in idealized form and then to adapt it with as little disruption as possible to the space actually available. There are two types of layout, i.e. in-line and return type.

The in-line layout involves feeding the work into one end of a line of tanks and unloading at the far end. Many hoist-operated plants work on this system. A disadvantage is that the carrier bars for the work (also known as flight bars) have to be returned over the line for reloading. In some plants the flight bars are returned outside the plant line area (see Figure 4).

FIGURE 4. IN-LINE HOIST-OPERATED PLANT FOR ITEMS UP TO 5 METRES IN LENGTH. FLIGHT BARS ARE RETURNED ON SEPARATE TRACK. THE MONO-RAIL FEEDING WORK INTO THE LINE IS SEEN AT THE FAR END. (Courtesy: Acorn Anodising Co. Ltd., Hayes, Middlesex.)

A return-type plant is common in small manual units and is also adopted for large-scale operations where it is desired to feed work in and out at the same end. A modern return or "horse-shoe" layout is

Anodizing Equipment 41

often characterized by the automatic transfer of work, in a rinse tank, from one arm of the horse-shoe to the other.

Although most anodizing plants consist of a series of metal or plastics tanks, several installations have been constructed in the form of a series of concrete tanks each lined with an appropriate material. This scheme minimizes the problem of protecting tank supports, etc., against corrosion. Such a layout is best chosen when a fixed series of processes is to be installed and where little or no alteration in requirements is anticipated in the future.

A recommended feature in all plants which are not purely manually operated is the provision of suitable gaps in the tank line where the work can be inspected before passing on to the next stage. This facility is particularly important after etching, chemical or electrobrightening. If the processes have produced a faulty finish it is pointless to continue with processing. For similar reasons inspection after anodizing and after any colouring process is advisable in order to avoid any wasted effort in sealing unsatisfactory coatings.

A common fault in anodizing plants is an insufficiency and inefficiency of rinsing between each stage of the process. As a result an excess amount of impurities is carried over from stage to stage and can lead to severe contamination. Two- or three-stage rinsing with a countercurrent flow of water will provide the most effective and economical system. Where space is limited, the effectiveness of single rinse tanks can be greatly improved by fitting a row of water-spray nozzles along one or both top edges of the tank. If the work is always of such a nature as to permit complete rinsing by spraying, then it is advisable to dispense with tanks full of rinse water and to use, instead, tanks with several rows of water jets fitted on the inside walls: the water supply to the jets can be conveniently turned on and off by a foot valve actuated by the plant operator or by the flight bar actuating a time switch when the work is in the tank. Top sprays should be used when work is withdrawn from the tank and the speed of withdrawal should allow time for adequate rinsing by the sprays.

Chapter 7

Jigging (Racking) Methods for Anodizing

The anodic coating on aluminium has electrical insulating properties even when it is being formed; in a conventional sulphuric acid solution operating at, say, 16 volts, about 13 volts drop occurs across the coating and only 3 volts from the outer surface of the coating through the electrolyte to the cathode. To ensure continuity of the anodizing process it is therefore essential to provide firm electrical contact between the jig holding the work and the workpiece itself.

The jigging material can be aluminium or titanium. The latter material is suitable for anodizing to coating thicknesses of up to about 12-15 μm but it is liable to lose electrical contact on thicker coatings unless heavy-pressure screw clamps are used. Titanium itself acquires a very thin anodic coating during the process and the colour of this coating varies from yellow to purple depending on the precise anodizing conditions. The thin coating does not have to be stripped after each anodizing operation — it is sufficiently fragile to break when the work being jigged is pressed against it. Some typical titanium racks are shown in Figure 5.

In designing racks due note must be taken of their current-carrying capacity which must be sufficient to provide enough current for the total area of work on the fully loaded rack. The cross sections of titanium rack stems and wires should be based on a maximum current load of 0.2 amp per mm^2. The corresponding figure for aluminium is 1 amp per mm^2. These figures apply to titanium and aluminium in air, whilst when immersed in an electrolyte they can be increased by a factor of 5.

Titanium is much more expensive than aluminium and its use is therefore restricted to use on long runs of work or for making adjustable racks that can cope with shorter runs of a variety of sizes.

Jigging Methods for Anodizing

Articles must be suspended in the anodizing baths in such a manner that (a) all surfaces to be treated are exposed to the solution, (b) good, tight electrical contact is made with the conductor leading from the source of current, (c) jigs or racks neither contaminate the

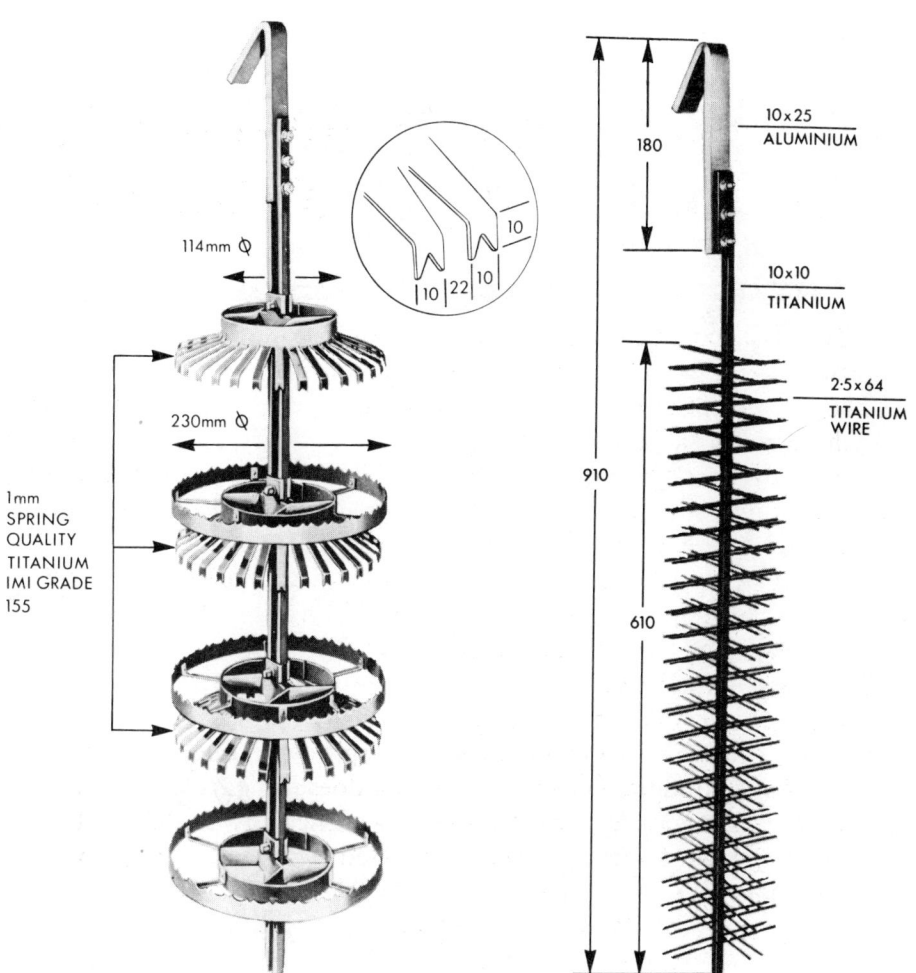

FIGURE 5. FOUR TYPICAL TITANIUM RACKS FOR ANODIZING SMALL ARTICLES (SEE ALSO OVERLEAF). THREE OF THESE ARE ADJUSTABLE TO ACCOMMODATE VARYING SIZES OF WORK. (Courtesy: Bunting Electrical Co. Ltd., Birmingham.)

solution nor become corroded by it, and (d) circulation of the solution is not impeded. Similar requirements apply to brightening and polishing baths, and it is common practice to rack articles for

44 Anodic Oxidation of Aluminium and Its Alloys

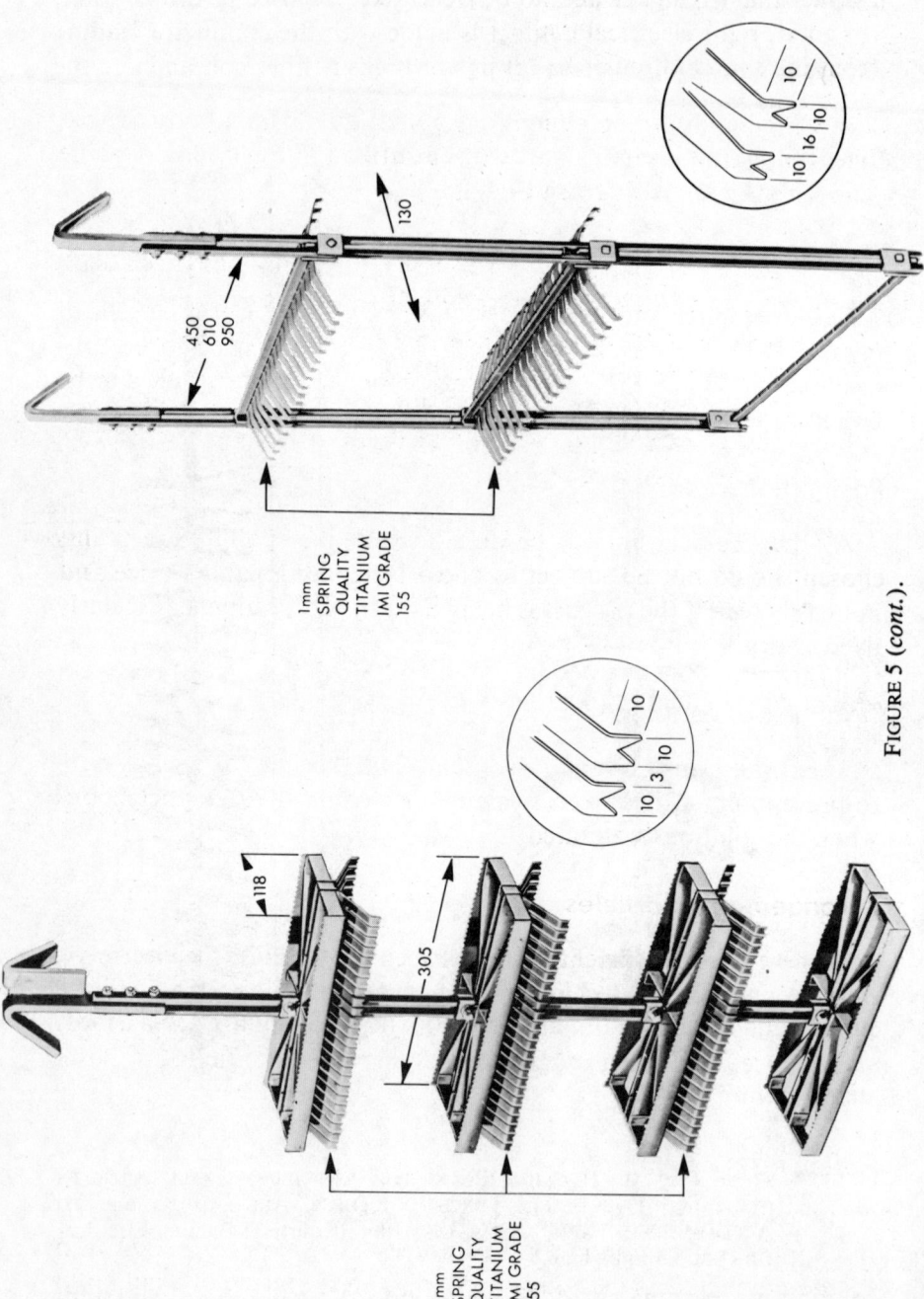

FIGURE 5 (cont.).

brightening as well as anodizing without intermediate re-racking. The following general rules should be observed.

Alloys

In general, different aluminium alloys should not be processed together, as the current densities will differ for each alloy and the final film thickness can vary widely.

Area of contact

This area must be as small as possible but sufficient for current to pass without overheating the metal. Large articles need to be connected to the source of current at several points; they may need to be bolted on to jigs.

Position of contact

As the anodic film will be absent at contact points, the points chosen should not be subject to corrosive conditions in service and not easily seen if the process is being used for decorative, particularly dyed, work.

Tightness of contact

The contact must be good electrically and must be sufficiently rigid to prevent any article from swinging into contact with its neighbour when the solution is agitated.

Arrangement of articles

If there is not sufficient space between the articles, the electrolyte may become overheated locally and cause burning on the anodized surface, or a deposit of unreliable quality. All significant surfaces should face cathodes at roughly similar distances, normally not closer than 15 cm.

Hollowed components

Hollowed or recessed components should, as far as possible, be face upwards so as to prevent the formation of gas or air pockets. To ensure complete and even film formation on the inside of a hollowed article having a very narrow opening, an auxiliary cathode or cathodes may be suspended near or inside it in a suitable position and internal agitation provided. Conductors to such cathodes must be insulated (e.g. with glass or porcelain beads) from a possible short circuit with the anodic workpiece.

Other metals

Composite articles made of aluminium and other metals, except titanium, cannot be anodized unless the other metals can be stopped off.

A full range of stop-off lacquers resistant to pickling, etching and anodizing solutions is available commercially for coating the portions of articles not required to be anodized.

Racks and jigs

Racks and jigs must have good electrical conductivity, as they form part of the circuit carrying the anodizing current. Suitable metals — generally in the form of square or round rod or wire — include titanium, pure aluminium and the aluminium alloy 6063. Joints may be either riveted or welded.

The aluminium alloy 6063 in the TD condition makes good spring contacts. Expanded aluminium is often used for racks and baskets, and pure aluminium in the soft condition is used for twisting round articles. The twisting method is, however, wasteful in time, space and material and should only be used where jigging is unsuitable.

Electrical contacts with aluminium, that are insufficiently tight, are liable to become worse during anodizing (in contrast to electroplating), and a contact once broken cannot be re-made. Aluminium jigs and clamps acquire an anodic film which must be removed from the contact points before re-use. The best method of doing this is to dip the jig in phosphoric/chromic acid solution (page 97) or in caustic soda provided that there is no danger of the solution being retained in crevices. (Caustic soda is speedier in action but will dissolve the jig more rapidly.) Caustic soda stripping of sealed anodic films is rendered easier if the jigs are first soaked in 20% nitric acid for about 15 minutes. Mechanical methods of removing the anodic film include the use of a fine file, coarse abrasive paper, or a steel scratch-brush. To reduce the area that requires stripping after each cycle, aluminium jigs may be anodized and coated with "inert" material over the non-contact areas, but considerable skill and experience are then needed to strip effectively the exposed portions. Alternatively, the need for stripping may be avoided altogether by making the contact points of titanium, the aluminium portion of the jig being coated with a material such as PVC or chlorinated rubber.

Such a jig will last as long as the plastics coating remains intact, after which repairs to the joints between frame and tips are usually required, followed by re-coating of the frame. Composite jigs cannot be used throughout those anodizing cycles that include solutions having an effect on the plastics coating, for example trichlorethylene and the phosphoric acid based brightening solutions.

For long runs it is economic to make the jig entirely of titanium, welding being the most efficient form of construction. Titanium is stronger than mild steel, and with careful handling all-titanium jigs will last without maintenance as long as they are required.

In some cases, e.g. sheets to be anodized on both sides, only one row of work can be loaded into the bath resulting in an uneconomical poor amperage load. If this is a frequent problem it is better to use a wider anodizing bath and to arrange an extra row of cathodes down the centre. By this method two rows of work can safely be treated.

In recent years increasing use has been made of C-clamps made in plastics of a grade that will withstand cleaning, etching, anodizing, colouring and sealing. They are unsuitable for operation in chemical polishing solutions. The C-clamp holds the work firmly against an aluminium or titanium bar. They do not need stripping as their function is purely mechanical and they do not conduct any of the anodizing current. The bars against which the work is clamped will, if

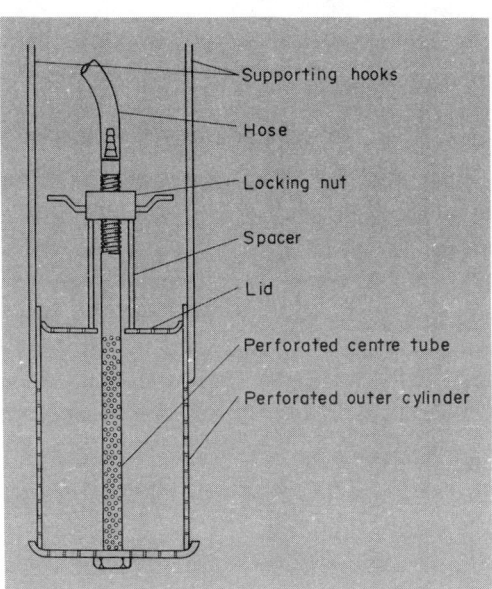

FIGURE 6. DIAGRAM OF BASKET ANODIZING SYSTEM FOR VERY SMALL COMPONENTS. (Courtesy: W. Canning & Co. Ltd., Birmingham.)

aluminium, need stripping after each load. Unlike racks carrying prongs at set distances apart, the C-clamp system provides adjustment.

Rivets and similar articles are very tightly packed in a perforated basket which can be plastics, aluminium or titanium, having a central aluminium tube (see Figure 6). This method involves a certain proportion of rejects, due chiefly to inadequate coating at the points where the articles touch one another or where the electrical contact is poor. Nevertheless a very high proportion of the articles are treated satisfactorily, rejects in one commercial plant ranging between 0.2% and 6% only. Articles which "nest" give most rejects, while irregular shapes (providing minimum opportunities for contact over substantial areas) such as bayonet lamp fittings commonly result in less than 0.5% rejects. Good electrolyte circulation through the tube is essential to avoid local overheating. Washers and other articles having flat surfaces cannot be processed in this way unless they can be mixed with irregularly shaped articles, but even then rejects occur.

Chapter 8

Chemical Treatment Processes before Anodizing

Removal of heavy oil and grease

Some parts become heavily coated with lubricant during fabrication. In order to avoid imposing too heavy a load on the wet cleaning processes it may be advisable to remove most of the oil in a vapour liquor apparatus using trichlorethylene or similar solvent. The articles can be degreased in bulk in baskets or after racking. When using chlorinated hydrocarbon solvent for aluminium the correct grade of solvent must be specified—it contains a stabilizer to prevent decomposition of the solvent by any accumulations of aluminium swarf, etc. This degreasing equipment is specially designed to minimize loss of solvent and to protect the operator against its anaesthetic effect.

An alternative method for dealing with this class of "soil" is to use an emulsion cleaner followed by vigorous spraying with water.

Chemical cleaning

The object of chemical cleaning is to remove polishing composition, oil, grease and general dirt from the aluminium so as to leave a clean surface ready for the following process. Most of these cleaners are mildly alkaline and formulated so as to produce either a slight etching action or none at all. Alkaline cleaners incorporating silicates as inhibitors are not favoured in view of the possibility of silicate being carried through to the sealing solutions. It is customary to buy proprietary cleaners which contain the necessary wetting agents, inhibitors, etc.

A useful general cleaner that will remove thin films of oxide (formed during heat treatment), polishing composition and grease is a 10% by volume solution of sulphuric acid at 90°C. In use the grease

50 Anodic Oxidation of Aluminium and Its Alloys

tends to float on the surface and is best removed by continuously pumping the solution over a weir at the end of the tank into an auxiliary tank. In turn the acid is pumped back to the main tank from the bottom of the auxiliary tank which thus acts as a grease trap. The grease is periodically removed from the surface of the grease trap. Hot sulphuric acid slowly attacks aluminium and this may be unacceptable on polished surfaces. The amount of attack can be minimized by adding an inhibitor such as aniline or O-toluidine, but the cost and toxicity of these must be taken into account. The slight spray involved when the solution is in production should be exhausted.

Alkaline etching

A solution of caustic soda (sodium hydroxide) at about 5% concentration is the cheapest and most widely used material for producing a clean whitish etch on aluminium. An operating temperature of 40-50°C is normal—the rate of chemical attack increases rapidly with temperature. Thermostatic control of the temperature assists in the production of a uniform rate of etching. The reaction between aluminium and caustic soda is exothermic so that heavily loaded solutions may need a cooling coil or panel to prevent a rise in temperature.

Although a plain caustic soda solution produces an acceptable etch its continued use results in the formation of rocklike deposits of aluminium hydroxide on the coils and tank walls. This defect can be overcome by adding one of a wide range of addition agents such as sodium gluconate, sodium tartrate or sodium heptonate. These are added at the rate of 3% of the weight of caustic soda used for the original make-up and maintenance additions. The insoluble aluminium hydroxide forms a soft sludge that is easily pumped away.

The rate of etching attack and, to some extent, the appearance of the finish can be modified by adding nitrates, nitrites, fluorides and other chemicals. These form the basis of proprietary etching mixtures. The best mixtures have good levelling properties, i.e. they flatten extrusion die lines with a minimum loss of metal.

The alkaline etching solutions do not dissolve all the constituents found in aluminium alloys, and the undissolved materials form a "smut" on the metal surface. Copper-containing alloys such as 2011 give a black "smut" whilst high silicon alloys (e.g. 4047A) produce a slate grey "smut" that does not dissolve in nitric acid. It can be cleared in a mixture of nitric and hydrofluoric acid. It is advisable and customary to "de-smut" the etched surface using either 10-30%

Chemical Treatment Processes before Anodizing

by volume nitric acid at room temperature or proprietary solutions based on ferric sulphate solution and nitric acid. In order to reduce the rate of neutralization of the "de-smut" solution by dragged-in alkali a two-stage rinse after etching supplemented by top sprays is recommended. Work should not be left in the alkali rinse solutions where it is liable to corrode.

Apart from controlling the etch solution by chemical analysis (see Appendix II), it is useful to check the actual performance of the etch by treating test pieces for a standard time and determining the weight loss of the test piece and also comparing the etched finish with a desired standard.

Etching proceeds at different rates on different alloys, but by careful selection of the operating conditions an acceptable match can be achieved between some alloys, rolled sheet and extrusions.

Acid etching

Acid etching solutions are needed for some special applications. The production of a matt surface for lithographic plates and also capacitor foil involves electrolytic etching in a solution based mainly on hydrochloric acid.

Solutions containing up to 10% hydrofluoric acid or a mixture of the acid and 2% nitric acid are useful for etching alloys containing 5% or more silicon. A very mild acid etch suitable for removing heat-treatment scale contains 10% sulphuric acid and 2% sodium fluoride. All these acid etches operate at room temperature.

Electrolytic polishing

This section includes "electrobrightening" which also increases the surface reflectivity but removes less metal than electropolishing. The effect of both types of process is to bring about a micro-smoothing of the surface by preferential attack on the high spots whilst depressions are protected from electrolytic dissolution by the temporary formation of a viscous layer that impedes the flow of current.

The electrolytic polishing processes are mainly based on phosphoric acid solutions with additions of other acids such as chromic, nitric or sulphuric, as shown in Table 7, and will smooth a wide range of materials. Good agitation, preferably mechanical, is desirable to produce uniform polishing conditions at the metal surface.

These processes have continued in use for many years in spite of the high cost of drag-out and electricity. The viscous solutions give off very little spray at the cathodes so that the problem and cost of

TABLE 7 COMPOSITION OF SOME PHOSPHORIC ACID ELECTROPOLISHING SOLUTIONS
(%)

Phosphoric acid (s.g. 1.70) (% by vol.)	75-80	95	80	42	75
Sulphuric acid (s.g. 1.84) (% by vol.)	20-25	5	–	8	25
Chromic acid	–	20 g/l	–	–	–
Nitric acid (s.g. 1.42) (% by vol.)	0.1	–	–	–	–
n-Butyl alcohol	–	–	20	–	–
Ethylene glycol monoethyl ether	–	–	–	33	–
Water	–	–	–	17	–
Temperature	85-90°C	82°C	52°C	75°C	50-70°C
Current density (amp/dm^2)	10-15	7-10	3-10	3-5	–
Time	Usually 2-15 min				

exhausting the spray and scrubbing it, if necessary, are much less than in the case of the chemical brightening solutions, most of which emit large volumes of toxic nitrogen oxides.

Although electropolishing will produce a bright smooth finish the subsequent anodic coating will lose its specular reflection properties as the coating thickness is increased, except in the case of the special alloys that have been developed for "bright anodizing".

With each bath the operating conditions vary as the aluminium content increases. To maintain a given current density, either the applied voltage must be increased or the working temperature raised. If the current density is too high, orange-peel effects may occur. In many cases the aluminium content stabilizes due to drag-out losses. A 20-volt supply is generally adequate for the operation of these baths, but 24 volts is preferable.

In solutions containing chromic acid, the use of a porous pot containing concentrated phosphoric acid round the cathode can reduce the formation of useless trivalent chromium.

Electrolytic brightening dates from 1934 and was the first process to be developed for improving the reflectivity of sulphuric acid anodic coatings. Even on mechanically polished super purity (99.99% aluminium) and on high purity (99.8% aluminium) an anodic coating is somewhat cloudy, due apparently to the presence of polishing materials occluded in the metal surface. The effect of electrobrightening is to remove a thin layer of about 10 μm of metal whilst improving the micro smoothness of the surface which, when anodized, gives an almost perfectly transparent film.

Chemical Treatment Processes before Anodizing

The first process "Alzak" was developed in the U.S.A. and based on a 2.5% solution of pure hydrofluoboric acid at 30°C with an operating voltage of up to 30. This process is virtually obsolete. The electrobrightening process "Brytal" developed by the British Aluminium Company Ltd. has survived to this day, and although its application is mainly confined to super and high-purity aluminium this type of electrolyte has been modified so as to render it useful on some lower grades of aluminium.

The Brytal process comprises three stages:

1. a short pre-etching in the electrolyte without electric current,
2. the electrobrightening process,
3. the desmudging process.

The work which should already have been cleaned is immersed in the Brytal solution for about 30 seconds, during which chemical attack takes place on the aluminium with an evolution of hydrogen. The function of this stage is to ensure a clean uniformly wetted surface. The slight etching effect is removed by the subsequent electrobrightening. However, the electrobrightening treatment time can sometimes be shortened by omitting the initial etching stage—this can only be established by trial.

At the end of the etching period the current is switched on and the voltage raised to the operation figure of about 20. The precise voltage depends on the chemical composition of the solution, its temperature and on the type of work being processed. The process can be carried out either in a still bath with baffle plates arranged to minimize movement of the liquid by the hydrogen evolved at the cathodes, or with a moving anode bar oscillating vertically or horizontally at, say, 12 4-in. strokes per minute. Significant surfaces should, where practicable, face slightly downwards.

The current density settles down to about 1.5 amp/dm^2 in a still bath or 3 amp/dm^2 with work movement. A well-smoothed DC supply is required and DC dynamos are still used in some installations. Single-phase rectifiers are unsuitable. An average treatment time is 20 minutes.

The solution composition is as follows:

Trisodium phosphate, anhydrous	60 g/l
Sodium carbonate, anhydrous (soda-ash)	200 g/l
Temperature	80-90°C

Aluminium is dissolved in the solution during the process and forms a sludge which is best removed by decanting. Losses of carbonate and phosphate, ascertained by chemical analysis, are made up by additions of sodium carbonate and trisodium phosphate, and the total phosphate is increased to 80/110 g/l as the solution ages. Some operators add sodium hydroxide to maintain the effectiveness of aged solutions. Full operating instructions are available from the licensors of the process (British Aluminium Company Ltd.).

During electrobrightening a thin iridescent film of aluminium oxide forms on the metal surface and this is removed afterwards by immersing in the following desmudging solution:

Phosphoric acid s.g. 1.75	35 ml
Chromic acid	20 g
Water	to 1 litre
Temperature	95°C

This solution does not attack aluminium but dissolves the thin oxide film, leaving the surface ready for subsequent anodizing.

Chemical brightening

Chemical brightening acts in a similar manner to electropolishing but requires no electric current. This type of process has been widely adapted as a basis for "bright" anodizing. The first of these processes was invented in the U.K. and used the following mixture:

Phosphoric acid s.g. 1.75	60-75% by volume
Sulphuric acid s.g. 1.84	25-40% by volume

This solution is used at 95-100°C. The work should be clean and well drained or dry before immersion in this solution. Hydrogen is evolved during the reaction but the highly viscous solution prevents spray emission. After about 2-3 minutes' immersion (heavier pieces requiring extra time to attain the operating temperature) the work is removed and drained. A white foam forms on the metal during the draining period. The work is then plunged into a 1% solution of chromic acid. The treated metal has a bright diffused finish and is particularly useful for removing polishing "burns" and thereby giving an attractive uniform finish. The solution is expensive to make up and the drag-out losses are high, but exhaust- and fume-scrubbing facilities are not required. A good finish is produced on a wide range

Chemical Treatment Processes before Anodizing

of aluminium alloys with less than 2% silicon—the main application is on commercially pure aluminium sheet 1050A 1350 and extrusions in 6063 (see Tables 1, 2, and 3).

A much more widely adopted solution is based on a mixture of phosphoric acid and nitric acid together with additions of sulphuric acid, boric acid, salts of nickel or copper, wetting agents and anti-fuming agents such as ammonium sulphate. In the U.K. these are marketed as the "Phosbrite" solutions. Different mixtures are recommended for manual and automatic operation to take account of the dwell time during automatic transfer from the brightening solution to the following rinse tank. The work is best moved in the solution during the process, preferably by vertical oscillation. These solutions are used at 95-105°C and emit nitrogen peroxide, especially when the work is withdrawn from the solution. These fumes must be exhausted and scrubbed before discharging to the atmosphere. An immersion of 2 minutes is the average. The finish has a high specular reflectivity, especially when used on alloys based on 99.7% + aluminium (see Table 3).

The composition of these solutions is controlled by chemical analysis, particular attention being paid to the nitric acid content (usually about 5% by volume of nitric acid s.g. 1.42). Drag-out losses are made up by adding the ready mixed solution. To offset the high cost of the solution it is now possible to buy recovery units to reclaim phosphoric acid from the first rinse tank which is kept warm and static. This rinse is followed by a running water rinse and then by a dip in a de-smudging solution which may consist of either 20% by volume nitric acid s.g. 1.42 or 1% chromic acid or a proprietary mixture based on ferric sulphate. These de-smudging solutions, all of which are at room temperature, remove the thin copper-containing film that is left on the metal during chemical polishing. After de-smudging and rinsing the surface is ready for anodizing.

Chapter 9
Anodizing Processes

SULPHURIC ACID PROCESS

Standard processes

In the past the standard sulphuric acid processes have been known under a number of proprietary names (such as "Alumilite" and "Eloxal GS") but such names are becoming less widely used. The following notes apply generally to all methods. The general relationship between anodizing conditions and film preparation is given in Table 13 (see page 127).

Strength of acid

Acid strengths of 8-35% (by weight) are used commercially. Dissolution of aluminium increases with strength.

The films produced in strong solutions tend to be more porous, softer and more flexible than those produced in weak ones. A good general-purpose electrolyte is one having an acid strength of 15% by weight. Some confusion can arise, as strength may be specified by volume, weight or density. A conversion table for the most frequently quoted units is given in Appendix I. For a given acid concentration the density will change as the bath ages, due to the increase in aluminium content, and the voltage required to produce a chosen current density will also increase. The bath is made up by adding the acid to the water; the water should be of suitable purity for the production of the highest-quality films (see page 39).

Temperature

For ordinary processes the bath temperature ranges from about 18 to 25°C; for most purposes 20°C is appropriate. The films produced in cold electrolytes are harder and less absorbent than those produced in warm electrolytes. The process known as "hard anodizing" (using refrigerated electrolytes) is discussed on page 66.

Close temperature control is important, because the current density and the rate at which the film is dissolved by the solution alter markedly with temperature. These factors in turn affect the properties of the film and, in particular, the "blooming" of the film, which in some forms may not occur until a long period after anodizing. High temperatures tend to give powdery films. Automatic control (for example by thermostat) to within $\pm 1°C$ is essential for consistent results.

Very efficient agitation is needed to disperse the heat generated during the anodizing process. Local overheating increases current density (or reduces applied voltage if current density is kept constant) to give films of uneven and uncontrolled quality. For the highest-quality work there must be sufficient agitation to prevent the temperature at the work surface (measured by a thermometer against that surface) rising more than 0.5°C (1°F) above that of the bulk electrolyte. Too much emphasis cannot be placed on the vital importance of adequate agitation. Failure to observe this rule can result in coatings that pass the normal inspection tests but which will fail during subsequent exposure to weather with the formation of a white chalky finish. Figure 7 shows a typical relationship between voltage, current density, temperature and agitation. Agitation is usually by means of air bubbles rising from perforated compressed air pipes (air pressure 5-15 lb/in^2) near the bottom of the tank, by means of motor-driven paddles, or by pump circulation of the electrolyte. With the pneumatic method it is essential that the air from the compressor be free of entrained oil in order to avoid contamination of the anodizing bath.

For warming up the anodizing bath on cold mornings, steam coils may be fitted at the bottom of the anodizing tank.

Electrical requirements

Direct current from a rectifier is generally used; the supply available should be able to give, at 16 volts, a total current (in amperes) of at least 1.5 × maximum surface area in square decimetres to be anodized at one time. This will cater for most requirements; less will severely limit anodizing scope. To produce thick, weather-resistant anodic coatings, for which the final voltages are typically 18 (aluminium and many alloys) or 22 (aluminium-silicon-alloys), a 24-volt DC source will be required.

58 Anodic Oxidation of Aluminium and Its Alloys

Alternating current is occasionally used in some continuous strip and wire anodizing installations.

For DC operation the cathode is normally of lead or aluminium.

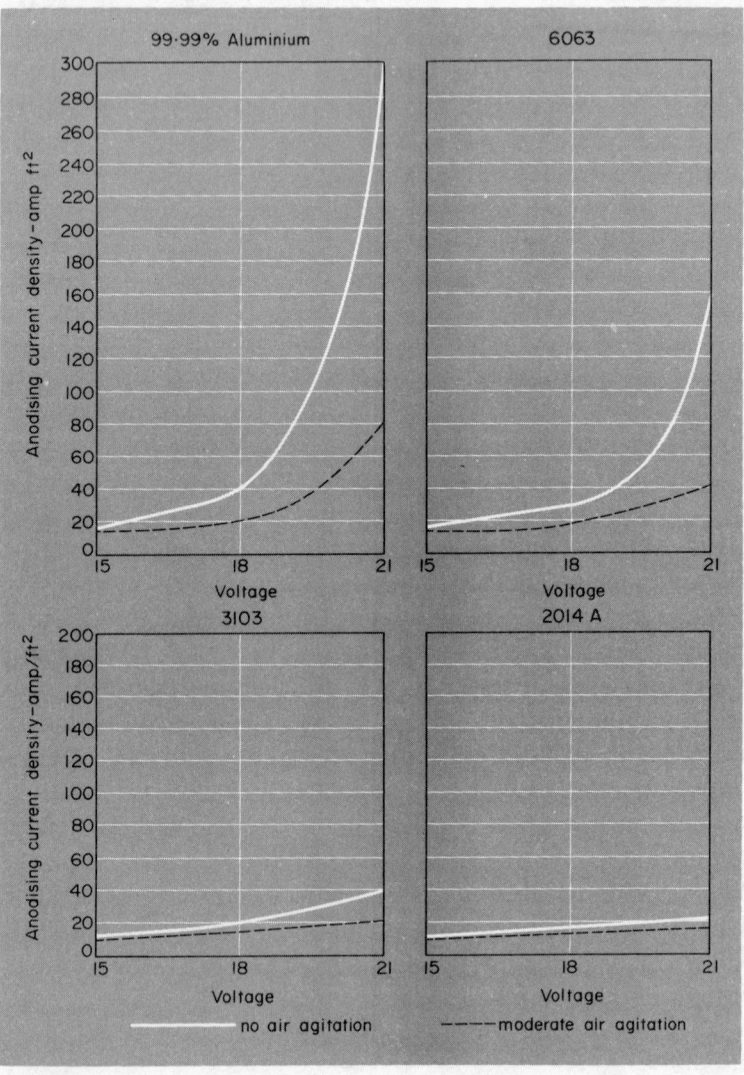

FIGURE 7. EFFECT OF INCREASING VOLTAGE ON THE ANODIZING CURRENT DENSITY, WITH AND WITHOUT MODERATE AIR AGITATION FOR 99.99% Al, 6063, 3103 AND 2014A. Spooner, R. C., *Metal Industry 1952,* **81**(13), 248-50.

Control of the anodizing process should aim at constant voltage. The resistance of the film increases as anodizing proceeds, so the current density decreases slightly during the process. For general purposes the current density should be in the range 1-1.5 amp/dm^2, but it can be as low as 0.5 or high as 3.5 amp/dm^2. The lower current densities are for thin, transparent films which are clearer than films of equal thickness produced at higher current densities; the higher densities are used in continuous anodizing. High current densities tend to counterbalance the high dissolution rates in strong electrolytes.

The voltage needed for any specified current density varies with the alloy and with the bath conditions but usually lies between 10 and 25 volts, the higher voltages applying to the lower acid concentrations and temperature. To provide the same current density when other conditions are equal, and compared with pure aluminium, the aluminium-zinc and aluminium-zinc-copper alloys (with or without magnesium) require a lower voltage, and the aluminium-silicon, aluminium-copper-magnesium and aluminium-manganese alloys need a higher voltage. Groups of alloys requiring different voltages cannot be anodized satisfactorily at the same time.

This operating voltage will also vary with the aluminium content of the electrolyte. It is good practice to choose a current density of say 1.5 amp/dm^2 and to treat a test load of the alloy that is to be treated. The surface area of this load plus that of the immersed racks must be carefully determined. A load of simple shape, such as a 75 × 6-mm section strip (in the case of extrusions) or sheets 1 × 1 m, is preferred because all the surface area is easily accessible to the anodizing current. The total load should be not less than 30% of a full bath load. The measured load is put into the anodizing bath and the current switched on. The voltage is raised until the calculated total amperage is reached. After a further 5 minutes the voltage is noted and the operating voltage is adopted for all work in the particular alloy for the day's work. This "calibration" process should be carried out at the beginning of each day's work, after any chemical additions have been made. Different alloys will need separate calibration loads. Where there is more than one anodizing tank, each tank should be separately calibrated. The "test load" can be stripped and re-used several times until the loss of metal area becomes significant.

Treatment period

The factors affecting the duration of anodizing are the thickness of the film that is required, the alloy, and the current density used. For pure aluminium, as a first approximation:

$$\text{Thickness } (\mu m) = \frac{\text{Current density amp/dm}^2 \times \text{Time (min)}}{3},$$

i.e. at 1.5 amp/dm² the film grows at 0.5 μm/min.

This rough guide must not supplant physical methods of measuring film thickness (see page 102). Considerable variations occur with extremes in processing conditions and with different alloys. The coating ratio* decreases with increase in temperature or in acid concentration (Figures 8 and 9) and is lower with most aluminium alloys than with pure metal. Films on aluminium-copper alloys may have less than half the thickness or coating ratio of those on pure aluminium for comparable anodizing conditions.

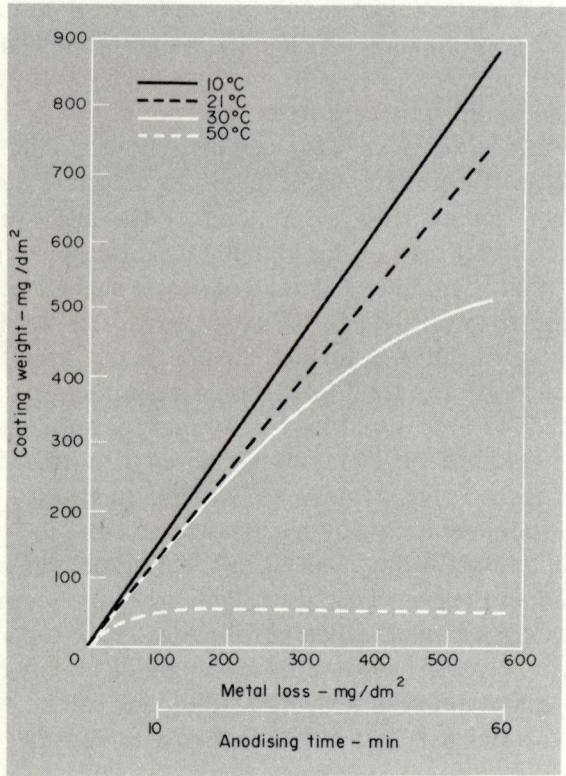

FIGURE 8. INFLUENCE OF ELECTROLYTE TEMPERATURE ON COATING WEIGHT/METAL LOSS RELATIONSHIP (15% SULPHURIC ACID, 19 AMP/FT²). Spooner, R. C. *J. Electrochem. Soc. 1955*, **102**(4), 156-62.

---

*The coating ratio is defined as the weight of anodic coating formed divided by the weight of metal removed. A low coating ratio indicates that much of the current passed is wasted in forming soluble products.

Operating practice

General practice should follow that customary in plating shops; the same Factory Acts are relevant. Operatives must wear protective clothing appropriate to the plant conditions.

When the bath is ready and the temperature checked, the parts to be anodized are lowered into it on their racks or jigs, these structures

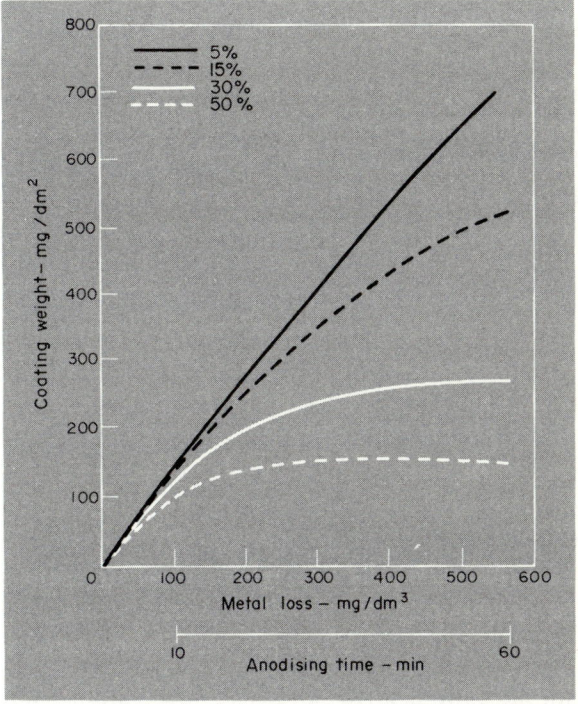

FIGURE 9. INFLUENCE OF SULPHURIC ACID CONCENTRATION ON COATING WEIGHT/ METAL LOSS RELATIONSHIP (AT 30°C, 16 AMP/FT^2). Spooner, R. C., *J. Electrochem. Soc., 1955,* **102**(4), 156-62.

being hooked or preferably clamped on to the work bars. Contacts must be tight, to avoid poor electrical contact (with consequent poor films and local burning) and to keep the articles rigid and clear of other work while the solution is being agitated. No articles should normally be closer than 15 cm to a cathode; large articles need more spacing — a large sheet should have a cathode at approximately the same distance on each side of it if films of even thickness are to be obtained. Thick gauge expanded aluminium is excellent as an auxiliary cathode. A cathode area equal to the anode area is usually good practice, but smaller cathode areas are satisfactory if evenly distributed. The total cross-section area of the cathodes must be sufficient to carry the maximum amperage loading of the system.

When all the connections have been made, the solution is agitated and the current switched on. The practice of removing smut from the articles in the sulphuric anodizing tank before switching on the current is not recommended for best results; it lowers the quality of the product and contaminates the solution, making necessary its more frequent renewal. On the other hand, work should not be put into the bath with full voltage on, as this will give rise to a very high current surge. To obtain films with maximum reflectivity the prescribed current density should be attained gradually over about 30 seconds; electrical conditions become stable after about 2 minutes. The voltage must not be raised above the required level and then reduced, as the barrier-layer effect will then prevent the full current from flowing. At the end of the anodizing period the current is switched off, and the parts are removed immediately and thoroughly rinsed in running water. (In commercial practice the bath is sometimes loaded and unloaded continuously, provided that excessive current surges are avoided.)

The anodizing solution has a good throwing power—very much better than plating solutions. With a good circulation of solution even the inside of a tube 30 cm long and 0.5 cm diameter can be given a coating which is not less than half the thickness of that formed on its outside.

Modifications to anodizing solutions

Mixed dilute sulphuric/oxalic acid electrolytes improve the efficiency of oxide film formation and its hardness. Good results are obtained with an electrolyte containing 120 g/l sulphuric acid + 50 g/l oxalic acid. The film produced in this mixture at 30°C is comparable with that produced in a 15% sulphuric acid electrolyte at 21°C (70°F). The mixed solution can be operated at about 10°C above that of a sulphuric acid electrolyte of equivalent acid concentration without loss of abrasion resistance—this is very advantageous in easing the problems of cooling, especially important in the tropics. About 0.2 g oxalic acid is consumed per ampere-hour.

Bath control

The preceding sections have emphasized the need for strict bath and operational control if reproducible results are to be obtained; in particular, close control of temperature ($\pm 1°C$), current density

(± 0.15 amp/dm^2) and acid concentration ($\pm 0.2\%$). Timing of bath operations also requires care: the current should be switched on immediately after immersion, and off immediately before removal of the work; for continuous working the articles are usually loaded and unloaded at set intervals.

It is important to keep the aluminium concentration at a constant low level — for 15% sulphuric acid the limits are preferable between 5 and 20 g/l; high acid concentration can least tolerate dissolved aluminium in the bath. A method of analysis for dissolved aluminium is given in Appendix II. The proportion of chlorides present should be well below the maximum of 0.02 g/l (as NaCl) specified in DEF151, while fluorides should not exceed 0.001 g/l (as F).

Contamination with phosphoric acid (often due to inadequate rinsing after chemical polishing) will result in coatings that cannot be properly sealed. A figure of 5 mg/l (as H_3PO_4) should not be exceeded.

The bath level falls continuously, due to evaporation, and demineralized water may be needed for topping up. For baths which are in regular use, a semi-continuous method of bath regeneration, replacing some solution every day, should be adopted. Equipment is now available for the continuous removal of dissolved aluminium.

Continuous Anodizing

Continuous processes are economically applied to the anodizing of long lengths of wire, strip and foil, the material being driven or pulled over a series of pulleys or rollers so that it passes through each treatment bath or rinse. The solutions and operating conditions are fundamentally as for ordinary sulphuric acid anodizing, but as the aluminium moves at a constant speed, the time of dwell in each bath is predetermined by the effective bath length.

The product is suitable for applications where electrical insulation requirements or the ability to take dyes is coupled with the need for some corrosion resistance and a ductile coating (Figure 10). It is used for electrical conductors and for mass-production lines where forming takes place after anodizing, as in the manufacture of radio trim, cans and bottle tops. The maximum anodic film thickness suitable for forming varies according to the processing conditions, but is usually about 2-6 μm. The anodic film is often much thinner; for example, canning strip has a film 0.2-0.4 μm thick as a base for lacquer.

Pretreatment

No mechanical pre-treatment is normally applied to wire and decorative strip, but in the continuous anodizing of strip for electrical insulation between turns of windings the low-voltage breakdown characteristics at the edges have been overcome by a mechanical or chemical polishing prior to anodizing. Chemical pre-treatments may include cleaning, etching or brightening, the solution concentrations being high to keep the time of treatment low. In one process, where strip is very thinly anodized, prior chemical cleaning is obviated by the addition of a wetting agent to the anodizing bath.

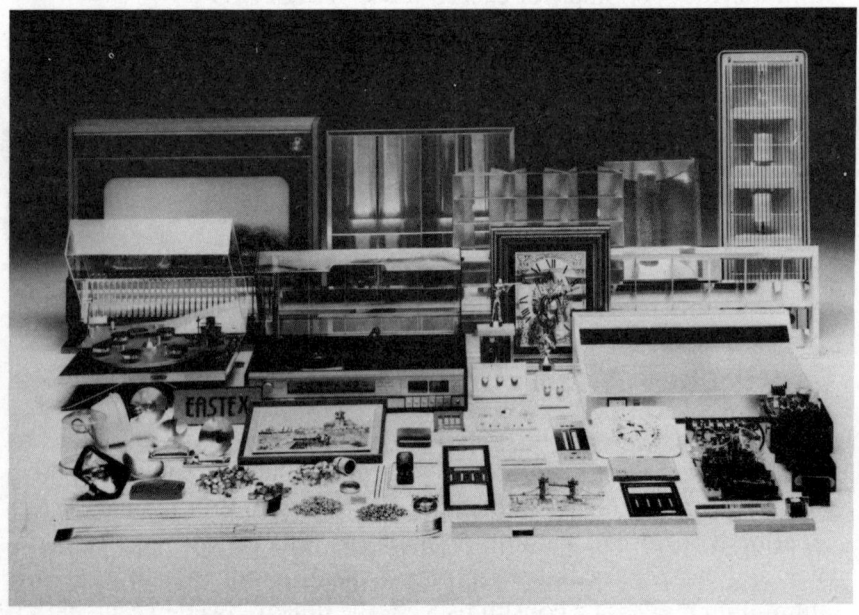

FIGURE 10. A TYPICAL RANGE OF ARTICLES PRODUCED FROM OR INCORPORATING CONTINUOUS ANODIZED SHEET. (Courtesy: Ano-Coil Ltd., Bletchley.)

Anodizing conditions

For wire, anodizing speeds generally in excess of 35 ft/min are necessary for economic reasons. To keep baths within reasonable size, this means that anodizing periods must not exceed about 45 seconds, and a film of 5 μm thickness will require a current much higher than the wire could carry in air. Several processes overcome this, using so-called liquid electrical contacts in which the current is conducted to the wire after it has entered the electrolyte in an entry chamber prior to passing into the main tank. Current densities are usually 30-90 amp/dm^2, either DC or AC; one commercial plant uses a combination of the two. When DC is used, the wire may be made cathodic while in the entry chamber and anodic as soon as it passes through a small nozzle into the main tank. With AC, two electrode chambers may be used, each being connected to one end of the secondary winding of the transformer; the electrodes consist of one or more pairs of wires equally distributed between the two chambers.

For the decorative anodizing of strip, slower speeds of 5-10 m/min obtain. While liquid contacts are used, in some cases ordinary contact rollers together with a mechanical shield at the point of entry into the solution are entirely satisfactory; if the voltage exceeds about 25 volts, however, sparking may occur where the strip enters the liquid and so cause pitting.

A high sulphuric acid concentration is desirable both for maximum conductivity (attained at 30% acid) and to give a flexible coating. The use of AC gives a more flexible coating than DC. These baths need cooling, and are usually worked above room temperature.

Wire from 0.0008 in. up to about 0.25 in. diameter can be processed, and most gauges and widths of strip. Separate plants are needed for materials of markedly different dimensions. Material processed must not be too soft, or it will stretch during processing; too hard a material will be difficult to coil. For many purposes 1200-H4 is suitable.

One of the principal continuous anodizing plants for strip for canning stock and building sheet operates at high speeds of say 100 m/min using a patented method consisting of anodizing with AC in a hot (90°) sulphuric acid solution at current densities in the region of 10-14 amp/dm^2. The time of anodizing is only 2.4 seconds. The very thin oxide coating formed is an excellent base for lacquering or for priming and painting (see Figure 11).

Treatment after anodizing

For strip this may follow the normal sequence, including dyeing (if required) and sealing, with modifications to suit the speed of the process. Thus a significant degree of sealing in boiling water may be obtained in as brief a period as 30 seconds, but longer or additional treatment is preferable. The speeds used for wire anodizing usually preclude such treatment, but the wound coils are sealed by immersion for 3-4 hours in boiling water, followed by drying at 82°C (180°F) for a similar period.

Anodic Oxidation of Aluminium and Its Alloys

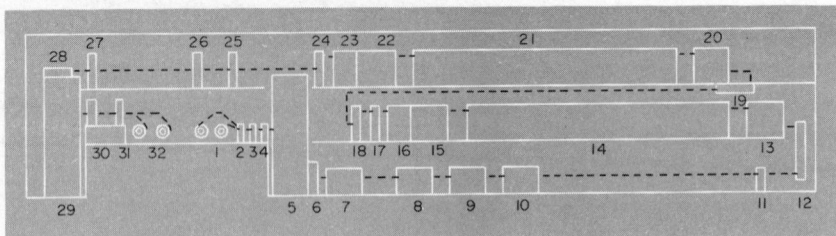

FIGURE 11. DIAGRAM OF CONTINUOUS HIGH-SPEED STRIP ANODIZING PLANT.
(Courtesy: Nordisk Aluminium AS, Oslo.)
A. **Entry section:** 1, Uncoilers. 2, Shear. 3, Joiner. 4, Bridle. 5, Accumulator. 6, Steering. 7, Tension leveller. 8, Anodizing. 9, Rinse. 10, Dryer. 11, Bridle.
B. **Coating section 1:** 12, Steering. 13, Coater. 14, Oven with high-temperature incinerator. 15, Air cooler. 16, Water cooler. 17, Steering. 18, Bridle.
C. **Coating section 2:** 19, Steering. 20, Coater. 21, Oven with high-temperature incinerator. 22, Air cooler. 23, Water cooler. 24, Steering. 25, Bridle.
D. **Exit section:** 26, Coating thickness measuring. 27, Strip thickness measuring. 28, Steering. 29, Accumulator. 30, Bridle. 31, Shear. 32, Coilers with belt wrappers.

High-speed anodized strip for lacquering is not sealed, but is dried and then lacquered and stoved. To protect the environment the fumes from the lacquering over are passed through an after-burner to destroy them. Alternatively, the strip is primer coated and then painted and stoved on a continuous line. This product is used for building cladding, caravan bodies, Venetian blinds, etc.

Hard Anodizing

Hard anodized aluminium may be defined as anodized aluminium with abrasion resistance as a primary characteristic, the production of which requires special anodizing techniques. The process of hard anodizing also offers means of building up worn or over-machined parts; coatings from 0.001 in. to 0.003 in. are widely used, but thicker coatings are possible on certain alloys. A dimensional increase on each surface of approximately half the coating thickness must usually be allowed for in engineering drawings. The hard anodized film is liable to craze on sharp edges; a radius of not less than 10 times the thickness of anodic coating is recommended to minimize this defect.

The solutions and operation of the process are fundamentally the same as for ordinary sulphuric acid anodizing, but all the variables are chosen to give a thick, hard, abrasion-resistant coating. DEF 151 and BS 5599 specify some properties of these coatings.

Hard anodizing impairs the surface finish; therefore surfaces to be

Anodizing Processes

treated should be smoother than required on the finished article unless a final smoothing operation is to follow anodizing. The process may reduce the fatigue strength of the aluminium alloy by up to 50% unless remedial measures are taken (page 68). The coating has poor thermal conductivity and does not perform well where frictional heat is produced. Its performance is also lowered under severe impingement conditions because it is supported on a comparatively soft base.

Pretreatment

A surface machined to about 4-8 μin. finish is usually suitable. Sharp edges should be radiused or chamfered to avoid crumbling of the coating; any drilling, straightening or bending of the components should be completed before anodizing.

Chemical cleaning is as given on page 49. Etching or brightening treatments are rarely used. If an article is to be anodized in some places and hard anodized elsewhere, the ordinary process should be completed first. The sections to be hard anodized are either stopped off meanwhile or are subsequently machined to remove the ordinary coating. If only a part of the surface is to be hard anodized the remaining surfaces must be stopped off with lacquer, wax or by mechanical means such as rubber washers or gaskets firmly fixed against the surface to be protected. The waxing, wax separation and cutting procedures used in hard chromium plating are equally suitable for hard anodizing.

Anodizing conditions

Concentrations of sulphuric acid of 15-35% by weight are favoured, and low operating temperatures (about −5 to +5°C) together with high current densities (2.5-15 amp/dm^2) give the desired film. To obtain these conditions a final voltage of 40-100 and a good refrigeration system are necessary. Many operating processes are patented. The "Hardas" process uses AC superimposed on DC, the proportions varying with the alloy being treated. Some alloys, notably those high in copper, require an initial AC "strike". The "Hiduran" process* uses DC at constant wattage, so that the required rate of heat extraction is constant throughout the process, minimizing the risk of "burning" at edges and corners. With this process it is necessary to use 30% w/w acid for high copper alloys such as 2014A.

*High Duty Alloys Ltd.

In America "Alumilite"* processes 225 and 226, which give a film thickness of 25 μm and 50 μm respectively on wrought materials, and processes 725 and 726, which give similar films on cast materials, are described as hard anodizing. They are operated at about 8°C (46°F) — a higher temperature than most processes but use sulphuric/oxalic acid mixtures. The "Martin Hard Coat" (MHC) process claims to use solid carbon dioxide in the electrolyte to provide cooling, and this may possibly have a specific effect on the coating. The current density is only 2-2.5 amp/dm^2, giving a slow rate of coating formation. The "Sanford"† process incorporates special organic additions to the anodizing bath, but the coating apparently has no unusual features. All these processes may encounter some difficulty with aluminium-copper alloys due to "burning" of the metal surface.

Thick, hard films can also be produced on some alloys in the integral colour electrolytes at room temperature.

Treatment after hard anodizing

The coatings are not normally sealed but are lightly waxed. In some cases they may be sealed in boiling water, dilute aqueous solutions, steam or wax, as with ordinary sulphuric acid films. Boiling for 15 minutes in 5% potassium dichromate eliminates some of the drop in fatigue properties resulting from hard anodizing. This treatment slightly reduces abrasion resistance.

CHROMIC ACID ANODIZING

The earliest commercial development of anodizing was with chromic acid electrolytes (by Bengough and Stuart) and their names are still widely used to describe chromic acid anodizing.

Strength of acid

The concentration of electrolyte ranges from 2% to about 15% (by weight) of chromic acid — i.e. chromic anhydride, CrO_3 — made up with pure water in a mild steel tank. In commercial practice two commonly used strengths are indicated by the familiar names of the processes concerned, i.e. "the 3% chromic acid" and the "10% chromic acid" processes. The first is a 2-5% acid bath conforming to DEF 151 for chromic acid anodizing (i.e. the original (Bengough-Stuart)), while the second uses acid strengths up to 10% to produce

*Aluminium Co. of America.
†Sanford Process Co. Ltd., Los Angeles.

thicker films. The maximum reproducible film thicknesses obtainable are about 8-10 μm on most alloys, but 6061 only produces about 5 μm and the aluminium-copper alloys about 2.5 μm (Table 8).

TABLE 8 THICKNESS OF ANODIC FILM ON DIFFERENT ALLOYS WITH VARIOUS CHROMIC ACID PROCESSES

Material	Film thickness (in μm) produced by			
	10% CrO_3 35 min 53-55°C 30 V	10% CrO_3 60 min 53-55°C 30 V	3% CrO_3 40 min 40°C* Voltage to DEF 151	3% CrO_3 60 min 40°C* 40 V
99.99% Al + 1¼% Mg	5.9	9.6	7.15	6.4
1080A-½H	6.35	9.45	7.75	6.2
1050A-½H	6.5	9.9	8.45	6.05
1200-½H	6.35	9.3	8.0	6.5
3103	6.35	9.0	6.5	5.75
5251-½H	7.3	10.3	8.0	6.2
21014A	2.4	2.6	1.5	1.6
6061	4.8	5.4	5.6	4.9
6082	6.9	8.85	7.15	6.4

Data supplied by Alumilite & Alzak Ltd.

*30°C for 2014A.
The experimental error is of the order of ±7.5%.

Temperature

Low temperatures produce harder and more compact films than high temperatures. Highly alloyed materials may not form a protective oxide film if the temperature is too high. The normal operating temperature for the 3% chromic acid bath when processing wrought materials is 40°C (104°F) ± 2°C (4°F); this control will achieve a closely reproducible finish (e.g. where the films are to be dyed). Most cast alloys are more satisfactorily anodized at 25-30°C (77-86°F). The optimum temperature for the 10% chromic acid bath is 54°C (129°F) for both wrought and cast materials. Both solutions require steam-heating coils, water-cooling coils and automatic temperature control. Solution agitation is desirable but not as essential as in the sulphuric acid bath.

Electrical requirements

Normally a stainless-steel cathode is used, but part of the tank itself can form a cathode provided the total cathode area is not more than one-fifth or less than one-tenth the total anode area. Excessive cathode area causes premature deterioration of the bath by reduction of chromic acid to the trivalent state.

Film formation, as described already, is governed by current density, which is in turn governed by the applied voltage; as with the sulphuric acid process if constant current density is required this voltage must be increased as anodizing proceeds.

In the 3% chromic acid process four types of treatment cycle are operated (Figure 12). The short-cycle process is very economical where crack detection or the provision of a key base for painting are the main considerations: the film thickness is much less than that produced in the standard cycle, and the power and chemical costs are proportionally reduced. The electrical equipment must provide a current density of up to 0.5 amp/dm^2; the average current density is about 0.3 amp/dm^2.

In the 10% chromic acid process a constant operating voltage (usually 30 V) is maintained. However, alloys containing copper and/or zinc, as major additions, and most casting alloys require only 15-20 V to maintain a similar current density—about 1.2 amp/dm^2. A soft, thin coating is produced at higher voltages with these alloys.

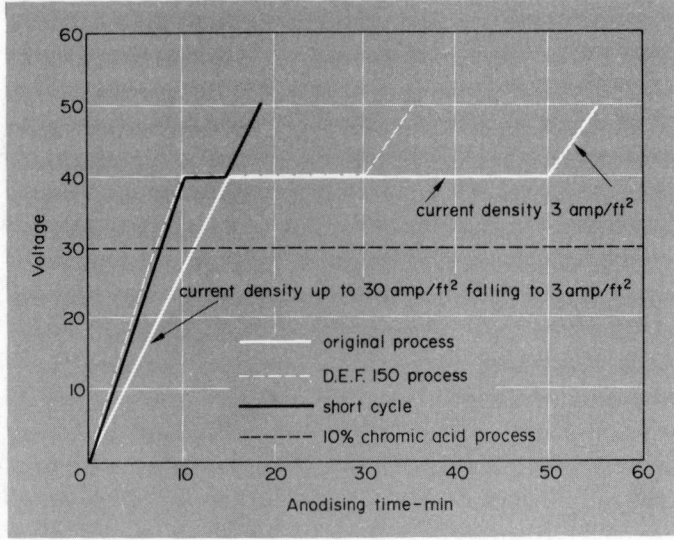

FIGURE 12. VOLTAGE VARIATIONS IN THE CHROMIC ACID PROCESSES (AT 40°C) ON COMMERCIAL PURITY ALUMINIUM.

Operating practice

The mechanics of operation are similar to those for the sulphuric acid processes. After anodizing and rinsing, a final rinse in demineralized water is useful to prevent drying stains.

The film is usually sealed as for sulphuric acid films (see page 89) but may also be treated by dipping in lanolin. Ten minutes in the sealing bath is adequate; a longer time may cause some film dissolution if the bath has become acidic. When the film is to be subsequently painted, it must be left grease free.

Chromic acid spray and fumes are harmful. The Factories Act, and particularly the Chromium Plating Regulations 1931, give stringent requirements for health and safety. The solution level should be at least 6 in. below the exhaust take-off. The use of special foam depressants stable in chromic acid greatly reduces or eliminates fumes.

The treatment programme laid down in DEF 151 is typical of U.K. practice. The work is hung in the anodizing bath at zero voltage and this is raised by steps of not more than 5 volts to a figure of 40 in 10 minutes. The 40 V is maintained for 20 minutes and is then raised to 50 V over a period of 5 minutes at which figure it is kept for a further 5 minutes. This completes the cycle. On cast alloys it is customary to stop the process at the end of the 40-volt cycle. Automatic programmes are available to carry out the above process, and in any case stepless regulators are preferred for controlling the rectifier voltage (Figure 13).

When the work is being anodized for crack detection only, it is rinsed quickly in cold water and allowed to dry. On other work there is always the possibility of acid seepage from overlapping joints, small bore holes, etc., and where the consequent orange/red staining is objectionable it can be minimized or eliminated by dipping in a 5% solution of ferrous sulphate crystals at room temperature, followed by rinsing, before proceeding to the next stages of dyeing or sealing.

Bath control

The aluminium content is usually kept below 0.3%. Chloride and sulphate contaminants in the bath are more harmful to the production of good-quality films. Although DEF 151 allows up to 0.02% chlorides (calculated as NaCl) and 0.05% sulphates (calculated as H_2SO_4), it is bad practice to work with more than a small fraction of these amounts of impurities.

British supplies of commercial chromic acid may contain up to

Figure 13. In Line Automatic Chromic Acid Anodizing Plant for Decorative Treatment of Small Articles. The Carrier on the Top Rail Conveys a Flight Bar of Articles through the Cleaning, Anodizing and Rinsing Processes. The Anodizing Voltage Sequence Is also Automatic. (Courtesy: Process Plant and Chemicals Ltd., Slough.)

0.2% sulphate, and most chromic acid electrolytes contain traces of it: this affects the degree of opacity of the film. Opacity is produced with as little as 0.02% sulphate present.

Chromic acid is lost from the solution both by drag-out and by reduction of chromium from the hexavalent to the trivalent state at the cathode during anodizing. In the trivalent state it is no longer effective for electrolysis. Additions of fresh chromic acid to make up for small losses must not exceed 200% of the original weight used in preparing the bath and should preferably be limited to 50-100%. Semi-continuous regeneration is preferred, similar to that described for the sulphuric acid processes. Cation exchange regeneration of chromic acid solution is more practicable and economic; it also keeps the aluminium and other metallic impurities at a low level.

Details of the method of analysis of the chromic acid content are given in Appendix II.

OTHER ANODIZING PROCESSES

Processes based on sulphuric and chromic acids have so many fields of application that in this country other processes are used only for special purposes. Oxalic acid processes give hard films with many properties comparable to those produced in sulphuric acid, but the cost tends to be greater; they can, however, be produced with a natural gold or straw tint. For electrolytic capacitors and for anodizing vacuum-deposited aluminium, barrier-layer electrolytes are used in which alumina is insoluble; examples are boric acid, ammonium tartrate, ammonium borate or ammonium dihydrogen phosphate. Other anodizing electrolytes include sulphosalicylic acid, which gives hard films coloured gold, brown or black; and phosphoric acid, which forms films suitable as a base for electroplating.

Brief notes on some of these processes follow.

Oxide-Solvent Electrolyte Processes

Oxalic acid

The solution normally used contains 3-5% oxalic acid, but up to 10% has been commercially used ("Alcanodox" process).* Operation is at 30-60 volts DC (German "Eloxal GX and GXL" processes), 40-60 volts AC ("Eloxal WX") or AC and DC ("Eloxal WGX"); alloys containing silicon or manganese require slightly higher voltages for comparable results. Temperatures are usually 15-35°C (59-95°F) and current densities 10-30 amp/ft^2. The current density is controlled by varying the voltage. The solution must be agitated vigorously in order that the characteristic light-straw colour may be uniform.

The process can yield very thick films if required, but for this purpose sulphuric acid films are normally preferred, unless there is a danger of acid being trapped in crevices. For electrical purposes, an oxalic acid film with the barrier layer thickened by subsequent boric acid anodizing is feasible.

Alternating current can also be used to produce films with brass yellow to bronze shades; the deeper colours are obtained on the thicker films. Alloys containing silicon always anodize grey with both AC and DC processes. The operation of oxalic acid anodizing and the control of the composition of the solution—in particular the avoidance of chlorides—are similar to those already described for sulphuric acid. Cathodes of carbon, lead, iron or stainless steel are suitable.

*Alcan International Ltd., Banbury.

After anodizing and rinsing, a 1% solution of ammonia or ammonium bicarbonate may be used to neutralize acid in joints and crevices. The film should be sealed in the same way as for sulphuric acid films.

Sulphosalicylic acid*

The recommended anodizing conditions are to use a solution of 40-100 g/l sulphosalicylic acid together with sulphuric acid equivalent to 30-60 g/l sulphuric acid at about 30°C with a current density in the range 2.5-3.5 amp/dm^2, which requires up to 80 volts. The time depends on the film thickness required.

Sulphophthalic and "Duranodic"†

This process operates in a similar manner to the sulphosalicylic acid method and provides the same range of hard coatings. A maximum voltage of 100 DC is required.

Phosphoric acid

The solution normally used contains 25-30% phosphoric acid at 25°C (77°F). Current density is 10-20 amp/ft^2 for 10-15 minutes, and the voltage is generally 30-60 V.

Compared with the sulphuric acid film, the practicable total thickness (up to 6 μm) is less. This is due to the greater solubility of aluminium oxide in phosphoric acid anodizing solution compared with sulphuric acid.

Barrier-Layer Processes

The films formed in barrier-layer electrolytes are much thinner than those formed in electrolytes which have some solvent action on the film as it forms. The use of distilled water is more important with non-solvent anodizing solutions than with solvent-type solutions. The film thickness depends almost entirely on the forming voltage, being approximately 13-14 Å per volt (10,000 Å = 1 μm). The choice of voltage depends on other factors also, such as the bath resistance and the need to avoid sparking. Adequate films are usually formed in about 45 minutes; the need for forming periods greater

---

*Kaiser Aluminium Inc.
†Aluminium Co. of America.

than twice this usually indicates defects or dirt in the metal surface, and the product is unlikely to make a satisfactory electrolytic capacitor. Unlike the films formed in solvent electrolytes, barrier-layer films do not require sealing—they are merely washed and dried.

Boric acid

Solutions of from 4% boric acid up to saturation are used, with additions of 0.05-0.5% borax. Avoidance of chloride impurity is important; as little as 0.01 ppm may prevent oxide formation. Temperatures range from 70°C to 100°C (158-212°F). The initial current density is typically 0.5-1.0 amp/dm^2, and this requires high voltages—from 50 V with the most concentrated solution to 750 V; the current then falls to a low level. The film, which is free from borates, has a thickness directly proportional to the final forming voltage.

Ammonium tartrate

These solutions normally contain 3% tartaric acid with ammonium hydroxide added to bring the pH to 5.0-5.5. The current density is initially about 2 amp/dm^2 but decreases in about a minute to a few mamp/dm^2, after which the film virtually ceases to grow.

Ammonium borate

With this solution, forming voltages of 600-700 V are used to give film having a breakdown voltage of 550 V.

Chapter 10

Colouring the Anodic Coating

The inventors of the Bengough-Stuart chromic acid anodizing process also discovered and patented the colouring of the coatings with organic dye solutions. Since that time this interesting and valuable property has been widely exploited. Coloured anodizing is used in many household articles, though often unrecognized by the general public. Apart from the production of coloured coatings by the integral colour methods already described, a number of techniques are available for colouring conventional sulphuric acid coatings.

Response of Anodic Film to Colouring

While natural anodic finishes are attractive in themselves, the range of decorative applications is extended by colouring the film; practically any variation in depth and hue can be obtained, as well as multi-coloured effects.

The main characteristics determining suitability for colouring are the thickness of the coating, its absorptive power and its natural colour.

On appropriate grades of aluminium the sulphuric acid process gives semi-transparent and colourless coating ideal for dyeing to pastel shades or full, rich tones. For maximum fastness to light, resistance to weathering and depth of colour, a 25-μm or thicker coating is essential. Where thin coatings are dyed – e.g. with bright anodized aluminium for artificial jewellery – the colour fastness is reduced.

The opaque off-white coatings of the chromic acid process are ideal for mellow pastel shades; the colour fastness is adequate for general interior work provided suitable dyestuffs are used, but is inadequate for external exposure.

With the oxalic acid process, effects may be obtained similar to those with sulphuric acid, but the shade is modified if the original oxalic acid film has a golden colour.

On silicon-containing alloys, all the usual anodizing processes give a grey or purplish-grey film which can be very attractive and is permanently colour fast, but it modifies the shade of dyed colours to a marked extent. Other tints due to alloying elements will also affect the final result.

With all anodizing processes, films with the maximum retention of colouring matter are obtained by anodizing at the higher limits of temperature, time and acid concentration. Other properties required in the anodic film, however, often modify these conditions.

To achieve uniformity of colour the anodic film must be of even thickness and be produced in conditions avoiding local variations in temperature or acid concentration.

Treatment before Colouring

After anodizing, work should be given two rinses in clean cold running water. Special care is needed to avoid the carry-over of acid into the dyebath. Dirt, grease and finger-marks must be scrupulously avoided, as they cause undyed patches in the finished work.

There should normally be the minimum of delay between rinsing and dyeing, to avoid lowering the absorptive capacity of the film for dyestuffs, however when, for example, stopping-off compounds must be applied for multi-coloured effects, the rinsing should be immediately followed by a thorough drying in warm air; incomplete drying will result in a patchy finish. Dried film may be stored for long periods, but with some loss in absorptive capacity.

The work should be wet when introduced into the dyebath, otherwise patchiness may result. A dry film may be re-activated by immersion for 1-10 minutes at room temperature in dilute nitric acid (5-30%), or in the anodizing electrolyte followed by the usual rinsing before colouring. These treatments increase the absorptive power of the film.

Selection of Colouring Agent

Organic dyestuffs and some inorganic pigments are used to colour anodic coatings. Dyes have different degrees of fastness, and for any given dye, the deeper the shade the faster the colour. Where fastness to light is important, dyes should be used which have a high fastness index, as assessed by BS 1615 or ISO 3843. In general, dyes with a value less than 5 are rarely satisfactory for general use, and a minimum value of 9 is specified for outdoor exposure.

Experience has shown that for maximum fastness when exposed to

Anodic Oxidation of Aluminium and Its Alloys

TABLE 9 ORGANIC DYESTUFFS FOR ANODIZED ALUMINIUM USED EXTERNALLY

Colour	Dyestuff	Concentration (g/l)	pH ±0.5	Temp. ±5°C	Time (min)
Yellow	Aluminium Yellow 3GL	3	5.5	60	20-30
Dark Blue	Aluminium Blue G	5	5.5	60	20-30
Turquoise Blue	Aluminium Turquoise PLW	12	5.0	60	20-30
Red	Aluminium Red B3LW	5	5.5	60	20-30
Black	Aluminium Deep Black MLW	8	4.5	60	30-40

All on a recommended coating thickness of 25 μm

sun and weather the suitability of dyestuffs cannot be determined by accelerated tests alone. External exposure tests for more than 15 years on a wide range of dyestuffs, many with colour fastness rating 8, have shown only those listed in Table 9 to have long-term colour fastness. Recommended processing conditions are given in the table. Further, maximum fastness to light of these dyes and pigments is obtained only when they are applied to an anodic film of appropriate thickness and quality, and with suitable processing conditions, these to be followed by two cold running water rinses in clean water for periods of at least 2 minutes. With all dyestuffs the colour fastness will be lower and the colour reproducibility less easily attained if dyeing is stopped before the full colour has developed. However, the resultant colour fastness is often adequate for interior work, and this limited time technique enables delicate colourings to be obtained which are particularly attractive and popular on anodized aluminium.

It is important to note that although an infinite variety of colours can be produced by mixing dyestuffs, the rate of absorption of each dyestuff component may vary, thus giving rise to a gradual change of shade. Furthermore, when dyestuffs having a high light fastness number are mixed the fastness of the new colour may be greatly inferior to that of the original components.

After removal from the dyebath the work is rinsed and then sealed as specified for the particular colouring treatment used.

For exposure to heat some dyestuffs and pigments are suitable up to 240°C, but many change colour above 100°C. The dyestuff supplier's advice should be obtained.

Most dyestuffs have some affinity for anodic coatings provided the correct techniques are used. For commercial work, hot or cold dyeing may be used, but hot dyeing is preferred. In many cases a simple dye solution gives the maximum affinity, but other dyes require a small addition of acetic acid to adjust to the pH value recommended by the dyestuff supplier.

Colouring with Organic Dyes

After weighing, the dye is made into a smooth paste with cold water and then dissolved in boiling water before it is put into the dyebath; complete solution is essential. Softened water is very desirable in all dyebaths, particularly when maximum colour fastness is required. Mains water in some cases contains appreciable quantities of calcium and magnesium which form insoluble compounds with the dyestuffs, thus leading to a loss of expensive dyestuffs.

For hot dyebaths where no specific directions are provided by the dye suppliers a temperature of 65°C and a period of 20 minutes are recommended. Chromic acid films normally adequately dyed in 5 minutes, although 15 minutes or more is necessary for black. In practice, selected acid wool dyes and direct dyes at concentrations between 0.1% and 1.0% are widely used, the higher concentration being used for deep colours, particularly black. The bath concentration should be checked frequently by colorimetric comparison with a standard. For colour matching in predetermined dyeing times, the bath temperature should be kept within $\pm 1°C$ ($\pm 2°F$) and the pH within ± 0.5. Regular pH control is essential by means of frequent tests on small samples and adjustments of the pH by the addition of either acetic acid or sodium hydroxide (or ammonia).

Recent work has demonstrated the importance of determining the actual weight of dyestuff absorbed by the coating, particularly in the case of black dyeing for outdoor use. The analytical method is available from the dyestuff suppliers. The results obtained are more meaningful than the results given by colorimetric analysis of the dye solution. Temperatures above 80°C (176°F) are normally to be avoided, because sealing begins in the dyebath and the absorption of dyestuff by the film decreases. Short dyeing times are often used for manual operation where the skill of the operator is important. With dyes that can be applied in cold solutions the degree of dyeing is more easily controlled as it is slower.

The dyebath must be kept free from oil films (which can be removed with tissue paper) and other foreign matter; in all respects the bath must be kept absolutely clean. Where stainless-steel tanks are suitable and in use there should be no metallic connection through to the workpiece, as this can set up a local electrolytic cell and cause small pits and incrustations to appear in the anodic film. It is good practice to provide an insulating frame on the top of dye-tank flanges to prevent this type of galvanic electrical leakage.

The work should be kept in motion in the bath, or the solution should be gently agitated.

Colouring with Inorganic Pigments

Many of the general operating recommendations given for organic dyes also apply to inorganic pigments. In these processes insoluble coloured inorganic substances are deposited within the pores of the film, and in general the colours produced are more stable to light and heat than organic colours.

Gold shades obtained by immersion in 8-10 g/l ferric ammonium oxalate at a temperature of 40-45°C and pH 5.0-5.5 and are the result of iron oxide depositing from the complex salt within the anodic film. During use and arising from photochemical action iron hydroxide precipitates from the solution, however, and must therefore be redissolved with 5 g/l oxalic acid; after settling for 2 hours at 60°C the pH is adjusted.

Pigment colours can be obtained by double decomposition, but in view of the production difficulties involved the only example of any importance is the gold, bronze and black given by the cobalt acetate-potassium permanganate method.

In this case two solutions are made up consisting respectively of

30 g/l potassium permanganate,
50 g/l cobalt acetate.

The freshly anodized work is put into the cobalt acetate solution at 50°C for 2 minutes, quickly rinsed and transferred to the potassium permanganate solution at 30°C for 2 minutes. A gold-brown cobalt/manganese oxide mixture is precipitated in the coating. By repeating the operation the depth of colour is increased and with a specially absorbent thick coating it is possible to achieve a black finish. The operations are carried out under manual control and skilled operators are necessary to produce consistent colour matching. The light to medium bronze shades given by this process can be closely matched by one of the electrolytic colouring solutions described in the next section.

Electrolytic Colouring

The principle of electrolytic colouring was described by Carboni in 1936, and others, but we are indebted to Professor Asada of Tokyo University for investigating this matter in detail and devising the very successful methods now used on a world-wide basis.

In general terms the process consists of submitting the freshly anodized work to electrolysis in a heavy metal salt solution using AC, DC or special wave forms. The anodic coating acts as the cathode whilst the anode is usually made of stainless steel. At the cathode, metal is deposited in a finely divided form at the bottom of the pores of the coating. Depending on the metal in the electrolyte a variety of colours can be developed, their depth increasing with the time of treatment. For example, solutions of nickel, tin or cobalt, or mixtures of them, give colours ranging from light beige through bronze to black. The light fastness is outstanding. Copper solutions give pink-maroon-black colours but their corrosion resistance and light fastness are less reliable. Dilute solutions of silver salts give gold and bronze colours.

The precise solution compositions, equipment and operating conditions are available under licence from various patent holders in this field. Among the process names commonly encountered are "Anolok" (based on "Asada"), "Metacolor" and "Euracolor".

Multi-Colour Effects

Both random patterns and exactly reproducible designs can be produced in multi-colour effects. For either, the sequence of dyeing is governed by the following factors: (a) inorganic pigments should in general be applied before organic dyes and (b) where several colours are used the paler ones should be applied first. The characteristics of individual dyestuffs may, however, make an alternative procedure desirable, for example if the dark dye occupies a large surface area.

The bath temperature should be as low as possible initially, not exceeding 40°C (104°F) except in the final dyeing; the film tends to seal at high temperatures. Many dyes normally used hot will dye in the cold, but their colour fastness may be lowered. Heating in order to dry—if used at all—should be gentle at the intermediate stages.

Random patterns

These give mottled effects which are very attractive and, incidentally, disguise surface defects. They are produced by the following four methods:

(a) Precipitating mechanically unstable films of pigments such as lead sulphide on the surface of the anodic film by double decomposition, e.g. by successive application of 10% lead acetate and, after a rinse, 10% ammonium hydrosulphide. A water spray is then used to break the film away locally and reveal undyed areas which can be subsequently dyed with an organic dyestuff. If required, further areas of lead sulphide can then be broken away, followed by another dyeing operation.
(b) Using the ability of 50% (by vol.) nitric acid to bleach many dyes. The bleached surface is able to take further dyes, and this is used to produce mottled effects. Sharply defined areas of colour are obtained if the surface is dry before applying the nitric acid; wet surfaces give diffuse boundaries.
(c) Applying dyestuffs to local areas of dry unsealed coatings. Concentrated dye solutions (e.g. 2%) are used, and these solutions can be thickened by the addition of either glycerine or 20% of the following mixture:

Indian corn starch	75 g
Tapioca starch	25 g
Gum tragacanth (8%)	225 g
Water to 1 litre	

The starch paste may also be used as a partial resist, so that the other portions can be dyed in a cold dyebath. After drying, the metal is steamed for 15 minutes at 100°C (212°F), rinsed and dried.

This method may be adapted for spraying techniques, including the production of pattern effects. These are achieved by spraying through wire mesh or perforated plates, using different colours in different directions with a spraying angle of 45°.

(d) Floating a film of oils or similar material appropriate as a resist on water, and drawing the anodized aluminium article through this layer to leave a pattern of resist on the metal surface which will prevent dye adhering to these points. Alternatively the resist, in suitable solvents, can be applied direct to the anodized aluminium surface.

Colouring the Anodic Coating 83

The accurate reproduction of designs

The methods used for this purpose include processes in which some form of resist is mechanically applied to make a pattern, and other processes where light-sensitive substances are used either for impregnation or for the resist. Direct printing, using thickened dyestuffs and a rubber stereo, is more akin to the process (c) described above for random effects.

Dyes can be confined to particular areas on an anodized film by printing the remaining areas with printer's ink — normally by offset lithography. If the metal has yet to be anodized, or the anodic film is to be stripped and re-formed, a combination of ink and asphalt can be used instead. Ink and asphalt resists are applied by printing with a slow-drying ink, covering it with asphalt powder which adheres to the inked area only, brushing off the excess, and then melting the asphalt by baking at about 200°C (392°F). Such a resist is particularly useful on bare aluminium and can be used in various permutations with anodizing, dyeing and etching solutions. Typical etching solutions are (a) 10% hydrofluoric acid, (b) 5% caustic soda and (c) acidified metal chloride solutions, the materials used including magnesium, zinc, ferric and cupric chlorides and hydrochloric and fluoboric acids. The asphalt is finally removed with warm paraffin.

Resists applied either by offset or silk-screen printing form the basis of much multi-colour anodizing. After anodizing, and rinsing, the metal is warm air-dried. The sequence of operations is as follows: Print with the resist the areas that are not to be dyed, dye the first colour, dry, print with resists the areas to remain the first colour, bleach the exposed areas, dye the second colour, and so on; the resist is finally dissolved off and the surface is sealed. The bleaching solutions vary with the dye, and the dyestuff supplier should be consulted; weak ammonia solution, cold 1-10% sodium hypochlorite, 20-50% nitric acid, 10% sulphuric acid are among those used.

A later addition to the multicolour process is the Aluprint method (Sandoz Ltd.). Special dyestuffs are dissolved in an organic solvent and applied by printing or silk screen. These colours can also be applied manually to produce "paintings" or designs on a personalized basis. This type of colour is unsuitable for outdoor exposure.

Treatment after Organic Dyeing

Surplus dyestuffs are first removed from work by rinsing in cold water; deep colours such as black may need two rinses flowing in countercurrent, so as to avoid carry over of dyestuffs into the sealing tanks. Except where the dyestuff supplier advises otherwise all dyed work should be fixed by immersion in the following solution:

Nickel acetate — 0.8% by weight,

used at a pH of 5.6-5.8 at a temperature of 80°C for 2 minutes. The pH should be adjusted with acetic acid or ammonia. Sulphuric acid should not be used and boric acid must be avoided as a buffering agent as it reduces the resistance of the film to attack by sulphur dioxide. For large volumes of this fixing solution demineralized water should be used for make up and for replacing evaporation losses. Some aluminium hydroxide accumulates in the solution and should be removed by continuous filtration.

Sulphate is leached from sulphuric acid anodic coatings, and eventually interferes with the efficiency of the fixing. A safe upper limit is 1 g/l H_2SO_4. Where the expense of recovering is warranted, for example when a relatively fresh solution is accidentally contaminated with sulphate, the sulphate can be removed by the addition of the calculated quantity of barium acetate. The insoluble barium sulphate is then removed by filtration or decantation.

Photographic Processes

Light-sensitive chemicals can also be absorbed into the pores of the anodic film. In general the photographic processes are used more for the reproduction of diagrams and scales than for half-tones, although half-tones are of course possible using the silver chloride process. By using glass negative very close dimensional tolerances can be maintained. The process has been used, for example, for slide rules.

The preferred material is aluminium with a purity of at least 99.5% and a good surface finish. A thick, absorbent coating is needed, such as that produced in 20% (by volume) sulphuric acid at 27°C (81°F) with a current density of 15-20 amp/ft^2 for 50 minutes.

Silver chloride processes

There are several industrial processes. A method of impregnating the anodic film with silver chloride is as follows:

1. Flood or immerse the surface, using a 2.5% ammonium chloride + 2% tartaric acid solution.
2. Remove surplus solution, using a sponge or cloth.

3. By swabbing or immersion, apply a solution of 2% silver nitrate + 0.008% nitric acid.
4. Wash (to remove soluble salts) and dry.

Insoluble silver chloride is left in the coating. Surfaces thus prepared are stable to storage for up to a year, and in some countries metal that has been treated in this way is available commercially. The sensitivity of the film may be increased by adding bromides and iodides to the first bath.

Exposure is by ordinary methods, using a vacuum printing frame and a 2-kW arc lamp. Normal developing, fixing, toning and reducing techniques can be used, and the film must be sealed.

For developing (under a red safelight), an acid physical developer is used. It consists of 40 ml Solution A + 5 ml Solution B + 7 ml acetic acid + 2 ml 3% silver nitrate + 3 ml water, mixed immediately before use.

Solution A		Solution B	
"Metol"	10 g	Silver nitrate	80 g
Hydroquinone	20 g	Citric acid	40 g
Citric acid	12 g	Nitric acid	40 ml
Water to 1 litre		Glacial acetic acid	50 ml
		Water to 1 litre	

Silver is deposited from the developer itself into the exposed parts of the plate, as insufficient silver salt is held by the anodic film.

Fixing is in standard "hypo" solution under a yellow safe-light. Toning by immersion in 1% gold chloride solution can give a nearly black colour. The film is still capable of being given a dyed background, the dye being fixed and the surface sealed as discussed in the appropriate sections. The final appearance is improved by buffing with a swansdown mop and a polishing compound.

Blueprints

For making blueprints, anodized aluminium is first treated in a solution of 15% potassium ferricyanide and 15% ferric ammonium citrate. After removing surplus fluid with a clean cloth, the plate is exposed under a negative in direct sunlight for 3-10 minutes and developed with dilute nitric acid (5-15 ml conc. acid per litre). No fixing is required. The iron gallate process used in photoduplicating is similar, the solution used for this purpose containing tartrate for sensitizing.

Colour printing

The three-colour printing process is applicable to anodized aluminium. The process, briefly, is as follows.

Negatives are produced using a process screen and green, blue and red colour filters. The green-filter negative is printed on a plate sensitized with silver, the image is toned with gold, and the plate is then heated, so producing the red image. The plate is next sensitized with iron salts and the blue-filter negative is printed; this, when heated, gives the yellow image. Finally, the plate is again sensitized with iron salts and the third negative is printed, which produces the blue image. As each negative has been photographed through a process screen (which produced the image in a series of minute dots), the superposition of the red, yellow and blue dots one upon another results in the infinite variety of colours which make up the complete picture. Pictures, photographs and drawings reproduced in this way are nearly indestructible, as they are resistant to corrosion and unaffected by organic solvents such as alcohol, ether, benzene and amyl acetate; moreover, they are almost fireproof, as the coating resists even higher temperatures than does the base aluminium.

For most purposes, however, the printing techniques previously described are easier to operate and give good results.

Dichromate processes

Techniques using photosensitive resists based on dichromate-albumin or dichromate-gelatine can replace the silver chloride process, with equal dimensional accuracy and a wider tolerance in the type of anodic film.

For dichromate-albumin the process is as follows:

1. Wash and dry the freshly formed film at a moderate temperature, e.g. 60°C (140°F).
2. Pour on to the surface an emulsion of ammonium dichromate and albumin; whirl off the excess and dry the remainder.
3. Expose to a carbon arc through a positive transparency; develop by washing away the unexposed water-soluble albumin with water.
4. While still wet, dye and seal these areas of the film.
5. Remove the insoluble albumin by buffing.

The above procedure may give a yellow coloration to the "white" areas, due to the absorption of dichromate into the anodic film. This

Colouring the Anodic Coating

can be obviated by a reversal of the above process: the anodized film is dyed first, and after exposure of the dichromate-albumin through a negative and removal of the water-soluble albumin the unwanted dyed areas are bleached.

The procedure for the dichromate-gelatine process is similar to the standard process given above. Yellowing of the "white" areas is avoided by interposing a non-sensitive gelatine layer between the anodic film and the light-sensitive layer.

Chapter 11
Sealing the Anodic Coating

Mention has already been made of the ability of the anodic coating to absorb dyestuffs. The coating can, indeed, be likened to hard brittle blotting paper, and that whilst controlled absorption of substances by the film is extremely valuable it is essential to prevent the adventitious absorption of substances that may be unwanted or deleterious. These preventative methods are known as sealing processes, and fall into two categories: physical sealing and chemical sealing, which may be either by hydration or by reaction with metal salts, which itself may include partial or complete hydration.

PHYSICAL SEALING

The early coatings produced in chromic acid were afterwards painted or coated with a solution of anhydrous lanoline in a mixture of solvent naphtha and white spirit. These substances provided a physical blocking of the coating pores and for this reason this process is called "physical sealing" (Figure 14). A wide range of organic materials have since been used for physical sealing. They have some advantages in special applications. For example, sealing in lubricating oil or graphite/oil suspensions provides a lubricating surface that has been used for pistons. Polytetrafluoroethylene/resin mixtures will form a "non-stick" layer having a low coefficient of friction. Silicone coatings have also been specified for such mundane applications as curtain rails. Sealing with lacquer for use in the canning industry is referred to on page 65. The electrophoretic deposition of clear lacquer on to anodic coatings up to 10 μm thick, followed by baking, is commonly practised in Japan (the "Honnylite" process). The widely used electrophoretic coating with pigmented paint has also been used on anodic coatings to provide either an undercoat or a finishing coat.

Physical sealing does not decrease the surface hardness or wear resistance of the coating—important in hard anodizing applications.

However, some physical sealing materials can be removed by solvents or prolonged weathering, leaving an imperfectly sealed coating with properties inferior to those intended. There are practical difficulties in applying some of the physical sealing agents on a large scale and for this reason the advent of chemical sealing, first proposed by the Japanese in 1926, was welcomed by industry.

FIGURE 14. AUTOMATIC PLANT FOR ANODIZING THE INTERIOR OF BEER BARRELS PRIOR TO LACQUERING. THE BARRELS ARE CHEMICALLY CLEANED AND THEN ANODIZED USING AN INTERNAL HOLLOW CATHODE THROUGH WHICH SULPHURIC ACID IS PUMPED TO FILL THE BARREL. ANODIZING IS FOLLOWED BY SPRAY RINSING WITH DEMINERALIZED WATER. THOROUGH DRYING IS FOLLOWED BY LACQUERING. (Courtesy: Grundy (Teddington) Ltd., Burton-on-Trent.)

CHEMICAL SEALING

Hydration sealing

This type of sealing is the most widely used in commerce. Originally the process consisted of treating the well-rinsed anodic coating in saturated steam at atmospheric pressure or in boiling water. Although the exact mechanism of the process is not completely understood it is known with certainty that the following processes occur:

1. Some of the electrolyte which is absorbed or loosely combined with the oxide coating is dissolved away but is not completely removed. The residual sulphate (as SO_4) in a sealed sulphuric acid coating is typically about 13% — the precise amount depends upon the coating thickness and the anodizing conditions.
2. The original unsealed coating is almost anhydrous, a figure of 0.5% water is usually quoted. During sealing this water figure increases rapidly at first and later much more slowly. A conventional water-sealed sulphuric acid coating contains about 8-13% water, but sealing in steam at atmospheric pressure provides less hydration.
3. The absorption of water leads to an increase in weight of the coating but a decrease in the apparent density (i.e. including the air filled pores) from 2.6 to 2.4 g/cm^3.
4. The clearly defined porous structure becomes less distinct as the pores become blocked, starting at the outer surface.
5. The original aluminium oxide is partly transformed into an hydrated oxide resembling Böhmite $Al_2O_3H_2O$.
6. The outer layers of the film are reduced in hardness and are more easily abraded.
7. A thin powdery deposit known as sealing bloom forms on the surface.

The effect of hot water or steam sealing is to reduce or eliminate the ability of the coating to absorb dyes and the blocking of the pores increases the corrosion resistance of the coating. The electrical insulation is increased and current carrying ability decreased. These changes in properties are the basis of test methods for sealing (see Chapter 13).

THE PRACTICE OF STEAM SEALING

There are certain attractions, at first sight, in using steam for sealing. By its very nature the sealing temperature is constant and the steam condensing on the work is the equivalent of distilled water in purity. However, the condensing water sometimes washes any soluble salts from the work rods and racks on to the work and thus cause unsightly staining. The cost of providing steam on a batch process system is considerable. This can be overcome by using an inverted bell filled with steam and raising the work load up into the bell on special elevators. This system avoids the complete loss of steam between each load that occurs when ordinary covered tanks are used as steam chambers.

Steam sealing is normally used saturated and at atmospheric pressure, but some installations consist of a very large chamber into which a half day's or day's production can be loaded. The chamber is then hermetically sealed and the work is steamed under pressure.

The main precaution to be taken is to thoroughly wash the work and racks to remove soluble salts before sealing. Difficulties have also been experienced due to "dirty" steam from a boiler that is priming. Any risk of carrying over impurities from the boiler can be avoided by using a separate steam generator to supply the sealing chamber.

The time for steam sealing is similar to that for hot-water sealing, i.e. about 2 minutes per micron of coating thickness. After sealing the work will already be dry and will have a whitish sealing bloom which can be removed by immersion for a few minutes in a 30-50% by volume solution of nitric acid at room temperature followed by rinsing and drying.

THE PRACTICE OF HOT-WATER SEALING

This is the most widely used sealing process, but its apparent simplicity is deceptive. The factors that must be taken into account in order to obtain the best results are as follows:

1. *Temperature*

Ideally the water should be simmering at 99-100°C but this is wasteful in energy. A temperature range of 96-99°C is acceptable. It is difficult to maintain a uniform temperature in large volumes of water and specialized thermostat systems are needed to impose this close temperature range. The introduction of large masses of metal will cause a drop in temperature below the permitted range—extra immersion time will be needed to allow the full sealing time at the correct temperature to be achieved.

The rate of water absorption falls rapidly with a drop in temperature; at a figure below about 80°C the formation of the Böhmite type compound $Al_2O_3H_2O$ occurs. There is difference of opinion on the precise nature and compositions of the two forms of hydrated coatings; suffice it to say that the coatings sealed at lower temperatures fail the acceptance tests for good sealing.

2. pH of the water

The pH of the sealing water is important. At a figure below about 5.4 the quality of sealing decreases. At figures in the alkaline range, i.e. above 7, there is danger of chemical attack on the coating. In commercial practice a pH in the range of 5.6-6.6 is used. A pH range ± 0.1 is usually selected, the actual choice depending on the experience of the operator and the precise type of coating being sealed. Two common ranges are 5.6-5.8 and 6.4-6.6. The control of pH presents serious technical difficulties because residual acid electrolyte in the anodic coating tends to bring about a continuous reduction of the pH value. This effect can be corrected by the addition of an alkali such as caustic soda or ammonia, but unless this is done continuously on an automatically controlled basis the chosen pH limits may well be exceeded. For this reason it is advisable to make a preliminary addition of 5-10 g/l of sodium or ammonium acetate to the sealing bath to act as a buffer. In addition the drag in of acid in the coating can be minimized by soaking the work in warm water before sealing.

If the pH rises beyond the chosen upper limit it can be corrected by adding sulphuric or acetic acid, preferably the latter. When adding acid or alkali to adjust the pH, the reagent should be diluted with water and well mixed in the sealing bath. An air agitation pipe for this purpose is recommended.

3. Quality of the sealing water

Before the invention of ion-exchange resins mains water was used for sealing. As would be expected this caused problems in some areas. For example, in the south of England where the water has a high temporary hardness a continuous precipitation of calcium carbonate (and sulphate) was produced in the sealing bath, which in turn deposited a "lime scum" on the work. To counteract this difficulty, additions of acetic acid were made to the sealing bath before it was heated up, thus keeping some of the calcium in solution as calcium acetate.

With the advent of ion-exchange resins there was a move, especially in the continent of Europe, towards the use of demineralized water from which virtually all the anionic and cationic impurities had been removed. Unfortunately, this state of high purity is impaired as soon as the first load of work is introduced into the demineralized water. This defect can be overcome by continuously pumping off the slightly impure water, cooling it to a temperature that can be tolerated by the ion-exchange resin and then reheating and returning to the sealing bath. Alternatively, the "impure" water has to be dumped and replaced at suitable intervals as indicated by routine tests of the sealing quality (see page 107).

Demineralizing plant is costly to purchase and to run, and it is therefore recommended that sealing tests should first be carried out on the mains water supply (with the pH suitably adjusted). It may give very satisfactory results. However, some water authorities draw their supplies from various sources, i.e. rivers, bore holes or springs, and the proportion taken from each source may be subject to seasonal variations that upset the sealing quality. A note on the objectionable impurities in the sealing water appears in page 39.

4. The time of sealing

The basis of sealing times varies in different countries. In the U.K. a time of 2 minutes per micron of coating thickness has given satisfactory results. Elsewhere, times of 3 or even 4 minutes per micron have been specified. Whichever time is chosen it is essential that the sealed coating should pass the well-established acceptance tests.

5. Sealing bloom

Hot-water sealing, whether in mains water or demineralized water, causes the formation of a bloom on the oxide coating. In small-scale operations this can be removed by wiping or lightly buffing, but on a large scale this would be too labour intensive. Two methods of dealing with the problem here are in common use:

1. The removal of bloom in an acid dip, usually nitric acid, followed by rinsing and drying.
2. The use of special additives to the sealing bath to minimize or prevent bloom formation.

The second method has attracted considerable attention and a number of proprietary "anti-smut" agents have been developed. Whilst many of these do, in fact, prevent bloom formation, the quality of the sealed coating leaves much to be desired. Furthermore, some of the addition agents cannot be readily analysed so that control of their rather critical concentration is difficult. This is a field in which further developments are expected.

SEALING WITH CHEMICAL ADDITIVES

In this class of sealing the anodic coating is immersed in solutions of chemicals that react with the aluminium oxide in such a way as to improve its resistance to corrosion and general chemical attack.

METAL SALT SEALING

The Practice of Nickel Acetate Sealing

The earliest examples of sealing salts were based on nickel acetate, cobalt acetate or mixtures of the two.

A solution that has stood the test of time is

Nickel acetate	8-10 g/l
pH	5.6-5.8
Temperature	80°C

The addition of boric acid as a buffer should be avoided as it reduces the performance of the coating when tested by, for example, the Kape method (see page 110). In "soft" water districts the solution can be made up and maintained with mains water, but where the water has an appreciable calcium and magnesium content it is better to replace drag-out and evaporation losses with demineralized water. During use nickel is precipitated in the coating as nickel hydroxide. This inhibits staining by dyestuffs, increases the resistance of the coating to alkaline attack and, if the coating has been dyed, increases the light fastness of most colours. For some dye colours the nickel acetate treatment is essential to achieve the maximum light fastness.

The time of immersion in nickel acetate is usually 2-3 minutes, after which the work is rinsed and finally sealed in hot water. It will be found that dyes are prevented from bleeding out at the hot-water stage—a sure sign that the "fixing" process has been successful.

With some dyestuffs it will be noticed that some of the colour bleeds out into the nickel solution and if carried to excess the colour can tint other work. Thorough rinsing of dyed coatings before nickel sealing will minimize this trouble, but it may be found worthwhile continuously to filter the sealing solution through a filter unit containing activated carbon which will absorb the unwanted colour.

Where large volumes of nickel acetate solution are used it pays to prolong its useful life, especially as it is sometimes expensive to dispose of spent solution. The justifiable expenditure on maintaining the solution will be governed by local circumstances. The following systems have been used alone or in combination.

Sealing the Anodic Coating

1. Periodic removal of accumulated sulphates by adding the calculated quantity of barium acetate and filtering off or decanting from the precipitated barium sulphate.
2. Continuous filtration through activated carbon to remove unwanted dye.
3. Continuous filtration through a filter pack to remove the floc suspension of aluminium hydroxide that accumulates in the solution.

Nickel sealing produces a smut on the surface of the work which adds to the bloom arising from the subsequent hot-water sealing. This smut can be removed by hand wiping, mechanical polishing with a swansdown mop or by barrel polishing (in the case of small items).

To reduce the time involved in removing the smut patented systems have been devised to dissolve the bloom in chemical solutions leaving a clean surface.

As an alternative to nickel acetate sealing followed by hot-water sealing some operators add nickel sulphate to the hot-water sealing bath. This practice leads to the formation of a heavy bloom and the system is falling into disuse, except where a bloom-dissolving process is also incorporated in the plant.

The Practice of Dichromate Sealing

The corrosion resistance of anodized aluminium, particularly in a marine environment, can be improved by "sealing" in a solution of a chromate or dichromate. The sodium salt is usually chosen as it is cheaper than the potassium or ammonium salt. These formulations are prescribed in the DEF 151 specification of the Ministry of Defence and are based on work originally carried out in the U.S.A. and the U.S.S.R.

1. Sodium dichromate   70-100 g
    Sodium carbonate   18 g
or Sodium hydroxide   13 g
    Water   1 litre
    pH   6.3-7.4
    Temperature not less than 96°C

The immersion time for this solution is 5-10 minutes which is insufficient to bring about full hydration sealing but an appreciable quantity of chromate is absorbed by the oxide coating giving a distinct yellow colour. The intensity of the yellow colouring increases with coating thickness, all other conditions being the same, so that this sealing process will detect items with insufficient coating (for example, those that have lost electrical contact during anodizing).

2. Sodium dichromate 40-60 g
   Water 1 litre
   pH 5.6-6.0
   Temperature not less than 96°C

With this formulation the time of immersion is about the same as the time for anodizing.

This results in a fair degree of hydration but not necessarily complete sealing. For this reason the referee acid attack sealing tests are not applied to items sealed by either of the above methods.

Chapter 12

Stripping Anodic Coatings and Dyes

STRIPPING ANODIC COATINGS

Defective anodic coatings cannot conveniently be touched up: stripping and re-anodizing are necessary. Brush anodizing of local areas has been proposed but is not widely practised.

Surfaces contaminated with heavy oil or grease must be degreased before being stripped. Vapour or vapour/liquor degreasing in organic solvents is particularly useful for this operation. Lacquer on films also strongly resists the action of stripping solutions and should be removed by a suitable organic solvent or other stripping system specified by the supplier.

Anodic coatings are usually stripped in solutions containing caustic alkali (see page 46), but as these also severely attack the basis metal they must be used carefully.

Where attack on the basis metal must be avoided the following solution is used:

(a) Phosphoric acid (s.g. 1.75)   35 ml $\Big\}$ 98-100°C
     Chromic acid (A.R. quality)   20 g   (208-212°F)
     Distilled water to 1 litre

(Some use is made of a solution double this strength)

The coating takes 1-10 minutes to strip, depending on its thickness. This solution is also used in the reference method of determining coating weight (see page 103). The use of the double-strength formula for stripping hard-anodized films from some copper-bearing aluminium alloys may leave a grey to black surface which can be confused with the original surface, i.e. it could be thought that no stripping had taken place.

Re-anodizing can follow directly after stripping if the surface finish is still acceptable, but if the stripped surface is to be mechanically re-polished it must first be dried.

The dimensions of any part stripped of its anodic coating will be reduced by about half to two-thirds of the thickness of the original film. Part of the loss will be regained on re-anodizing. By anodizing to a greater thickness, the dimensions of the original anodized surface may be restored.

REMOVAL OF DYE

Some dyes and pigments can be removed, without harming the anodic coating, by the use of dilute nitric or sulphuric acid. If these are unsuccessful, 50 g potassium permanganate + 100 ml nitric acid with 1 litre of water at 20-30°C (68-86°F) for 2-20 minutes may be effective. As a last resort, the surface may be immersed in ferric ammonium oxalate to obtain an exchange of dye radical, and then stripped in dilute nitric acid.

ISOLATING THE ANODIC COATING

Anodic film in isolation may be required for use as a membrane or for research purposes. The following method will readily give a "window" of oxide coating, up to 1 cm^2 in area, in small pieces of anodized sheet:

1. Stop off the edges of the specimen if not anodized.
2. Abrade away the anodic film from one surface over a central area of, say, 6 × 6 mm. A sharp scalpel is a suitable tool.
3. Immerse the specimen in 100 ml of pure dry methyl alcohol and add 2 ml of bromine. Gentle warming may be necessary before the solution reacts with the bared aluminium. Do not allow the reaction to become vigorous, as the required oxide coating may be disrupted.
4. Let the reaction proceed — adding further bromine as necessary — until only a "window" of oxide is left in the metal frame. This will take several hours.
5. Wash the specimen at least twice with pure methanol, and allow it to dry in air.

If the anodic coating is required without the metal frame, then of course all the oxide on one side of the specimen should be abraded

Stripping Anodic Coatings and Dyes 99

away. The unsupported coating is very fragile and requires delicate handling.

The coating can also be separated from the basis metal by removing a portion of the coating and then immersing the specimen in a 5% solution of mercuric chloride. The exposed aluminium surface reacts with the mercury salt producing a voluminous aluminium oxide. Eventually the metal is all converted to oxide and the anodic oxide coating floats off in flakes. These usually curl up due to the inherent stress in the coating.

A further method is applicable to a sulphuric acid anodic oxide coating at the time that it is formed. At the end of the anodizing operation the voltage is reduced during a period of not less than 1 minute, or at such a rate that the amperage falls steadily to zero. During this operation the barrier layer section of the coating is reduced in thickness. If the coating is now left in the acid, without applied current, the barrier layer is dissolved at a greater rate than the rest of the coating and the coating floats off the metal surface. A separate electrolyte for the coating-removal is used. This consists of a conventional sulphuric acid anodizing solution at 23°C connected to a 50-Hz AC supply. After reducing the voltage to zero as described above the specimen is transferred to the second solution for the separating process.

Chapter 13

Testing Anodized Aluminium

Visual assessment of the surface is only of limited value with anodized aluminium. Even when the alloy and the anodic coating thickness are standardized, other important properties—such as corrosion resistance and wear resistance—can vary widely due to different processing conditions. Many of the test methods that follow are already incorporated in BS 1615, BS 3986 and BS 5599 covering anodic oxidation coating for aluminium; others are being considered for possible future inclusion. BS 1615 is being revised and issued in parts. In this section a general description of the tests is given: reference to the list of specifications on page 163 will give sources of fuller details. Although laboratory facilities are needed for the complete testing of an anodic film, visual checks on appearance, colour and texture, together with instrumental measurement of the coating thickness, and tests for sealing quality, can be made on the spot.

APPEARANCE AND COLOUR

Absence of banding or streaking is important. Depending on the particular application, a viewing distance should be fixed at which the surface should appear uniform. Samples should be agreed in advance for reference as to maximum tolerances on texture and/or colour matching.

Instrumental methods are available for measuring and recording colours and colour differences, but in commercial practice it is preferred to rely on the human eye for colour-matching purposes. It is important that the method of viewing samples for matching should be standardized.

(a) The specimens to be compared should be held in the same plane.
(b) The viewer should stand with the back to the light (preferably north light), with the light falling perpendicularly on to the specimens.
(c) The specimens should be so held that the viewing is also perpendicular to the coloured surface.

Another requirement for good colour matching (often overlooked) is that the viewer should have successfully passed tests for freedom from colour blindness!

Work that is etched before anodizing should also be checked for appearance to ensure that a uniform degree of mattness is achieved. Here again the human eye is a reliable guide and the method of viewing is the same as for colour matching. Some variations in etching will be unavoidable, and if this variation is critical, for example where the work is to be assembled in adjacent areas, it is possible to grade the items using a simple reflectometer (see Figure 15).

It should be noted that colours which match in artificial light may not match in daylight, and vice versa.

FIGURE 15. GLOSS METER FOR MEASURING HIGH GLOSS OR DIFFUSING SURFACES. (Courtesy: Sheen Instruments Ltd., Sheen, Surrey.)

COATING THICKNESS

Microscopic section method

A number of ingenious methods have been devised to measure anodic oxide coatings in the range of up to 5 μm and above. The reference method is based on preparing a microsection of a film of 5 μm or greater and measuring this under a microscope fitted with a graticule. From the known magnification of the system the thickness of the coating can be calculated.

The preparation of good microsections calls for skill and experience and it can be used to prepare test pieces with a known film thickness for the calibration of other types of film measuring equipment. It is useful to prepare test samples of about 10, 20 and 25 μm. The regularity of the boundary between the coating and the basis metal also provides useful information on the uniformity of coating thickness which may vary considerably on some alloys due to a differential rate of dissolution and oxidation of intermetallic compounds in the alloy.

Split-beam microscope method

Coatings that are reasonably transparent and where the basis metal is not unduly etched can be measured by an optical method whereby distance between the images reflected at 45° from the coating surface and the basis metal surface is viewed and the true coating thickness calculated from the formula—

$$T = T_1\sqrt{(2n^2-1)}$$

where T = the coating thickness,
T_1 = the apparent viewed thickness,
n = the reflective index of the coating which varies from 1.59 to 1.62.

Within an accuracy of 5%, $T = 2 \times T_1$ (see Figures 16 and 17).

The split-beam microscope is an elegant but expensive instrument. It is particularly useful for the rapid determination of coating thicknesses on bright anodized components. On etched samples the image reflected from the metal surface is diffused and if the etching is heavy this image can no longer be accurately located in the measuring eyepiece of the microscope.

Gravimetric determination of coating mass and thickness

This is one of the earliest methods for ascertaining coating thicknesses. Its success depends on the availability of solutions that will dissolve the coating without attacking the basis metal. The method can be used on all coating thicknesses and is particularly useful on coatings of 5 μm or less.

A measured area of coated material, free from grease, etc., is immersed in the following solution made up with distilled or purified water:

Phosphoric acid (d = 1.75) 35 ml/l
Chromic acid, CrO3 (reagent quality) 20 g/l

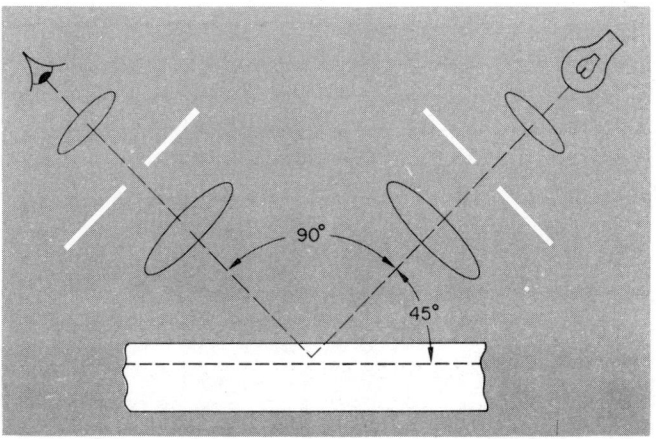

FIGURE 16. DIAGRAM SHOWING PRINCIPLE OF SPLIT-BEAM MICROSCOPE METHOD OF THICKNESS TESTING. (Courtesy: British Standards Institution.)

This is used at the boiling point and the coating dissolves rapidly. A whitish appearance on withdrawing from the solution indicates that the film removal is incomplete and the process must be continued until a clean metallic surface is revealed. The specimen is weighed before and after coating removal. The loss in weight divided by the area provides a figure for the mass of coating per unit area (a figure often specified in the U.S.A.). The density of the coating is about 2.4 for unsealed coatings and 2.6 for coatings sealed by hydration. For convenience these figures have been universally adopted. If greater accuracy is required the apparent density (including any air filled pores) can be determined by carrying out a weight loss determination in a coating that has had its thickness measured by another method.

From the weight loss measured above the coating thickness can be calculated from the formula—

$$T = \frac{1000 W}{ad}$$

where T = coating thickness in microns,
 W = mass of coating in milligrams,
 a = surface area of the coating in square millimetres,
 d = density of the coating.

Apart from the appearance of the specimen after film removal, the completion of the stripping operation can be checked by repeating the immersion in phosphoric/chromic acid for another minute followed by drying and re-weighing. No further loss of weight shall occur. On some alloys, particularly those containing zinc, a constant weight is not achieved due to continuing slow attack by the acid on the basis metal. In this case the original immersion time should be restricted to the shortest time for the visible removal of the coating.

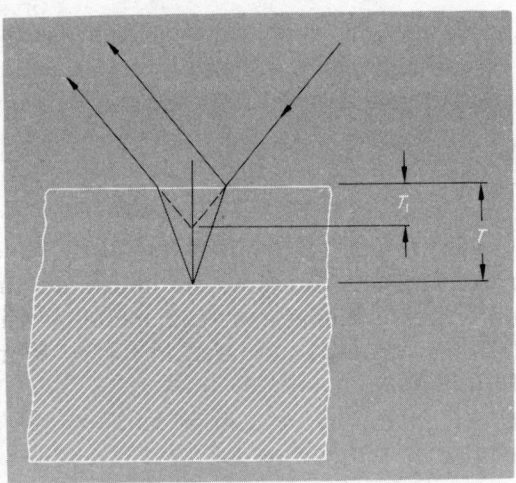

FIGURE 17. SPLIT-BEAM MICROSCOPE. RELATIONSHIP BETWEEN THICKNESS T AND THE APPARENT THICKNESS T_1. (Courtesy: British Standards Institution.)

Eddy current method

This is a non-destruction test that has been widely adopted for production-control purposes. The eddy current instrument has a probe in which a 1 kHz oscillation is produced. When the probe is applied to a metal surface it induces a current in the metal. The strength of this current is reduced if the probe is separated from the metal and this loss of strength is related to the distance of separation. The induced current is picked up by a detector in the probe and fed back to the instrument. In practice the induced current is converted to a scale showing coating thicknesses (i.e. distance of separation) in microns (see Figure 18).

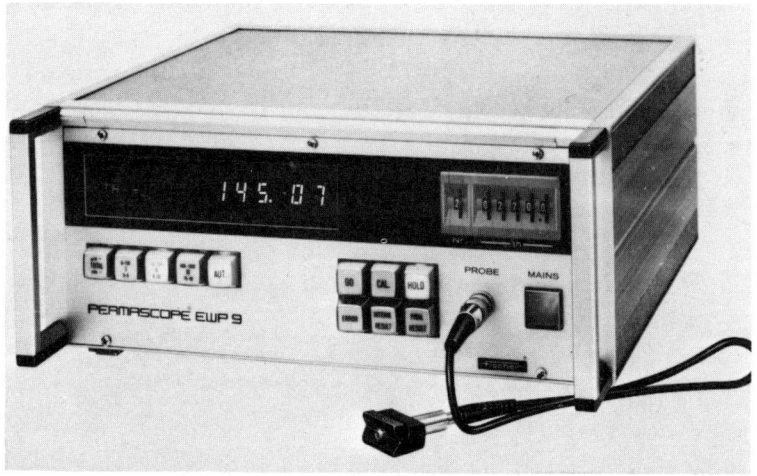

FIGURE 18. EDDY CURRENT THICKNESS TESTING METER WITH DIGITAL READ-OUT. (Courtesy: Fischer Instrumentation (G.B.) Ltd., Newbury.)

This type of instrument is inclined to "drift" and should be frequently checked on test pieces with a known coating thickness. The calibration process includes a zeroing operation using a piece of uncoated metal which should be identical in composition with that of the work to be tested.

An accuracy of about ± 1 μm can be achieved by this method. Sandblasted or deeply etched surfaces may give misleading results, but this danger can be minimized by using calibration samples, both coated and uncoated, that have been similarly sandblasted or etched.

Beta backscatter method

An elegant non-destructive method for measuring thicknesses of coatings, including anodic oxide coatings, is based on comparing the reflection of B-particles reflected from the top of the coating and the basis material.

106 Anodic Oxidation of Aluminium and Its Alloys

A source of beta particles A (Figure 19) is located so that a collimated beam of these particles is directed through an aperture B on to the coated component to be measured C. A proportion of these particles is returned back from the coating through the aperture to penetrate the very thin window of a special Geiger-Müller Tube D. The gas of the GM tube ionizes causing a momentary discharge across the GM tube electrodes; this discharge in the form of a pulse is counted by an electronic counter. Theoretically the GM tube pulses for each B-particle received, but in practice the efficiency is far less.

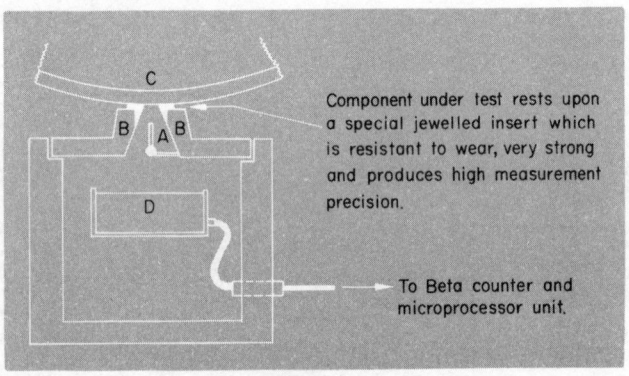

FIGURE 19. (Courtesy: Fischer Instrumentation (G.B.) Ltd., Newbury.)

If the beta backscatter received from both a sample of the substrate material and then separately from a piece of the coating material is known then the micro-processor, of the modern instrument, can be made to transform the number of beta particles received from a coated component into a meaningful indication of thickness, or weight per unit area.

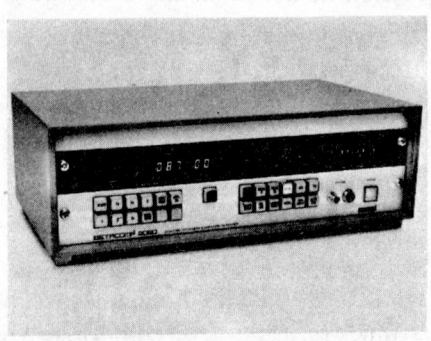

FIGURE 20. A MODERN BETA BACKSCATTER INSTRUMENT. (Courtesy: Fischer Instrumentation (G.B.) Ltd., Newbury.)

Figure 20 shows one of the most modern instruments in use today. Not only does it provide the digital indication of thickness, but it collects all the data received and provides real and relevant information on the quality of the products tested and their uniformity, then also at a touch of a button provides the user with the details of the instrument performance and the operator's calibration precision. Since permanent memories can store the relevant details for multiple measurement applications, and all results can be printed out, process control is made much more easy and cost effective.

ASSESSMENT OF SEALING AND COATING QUALITY

No aspect of anodizing has invoked more discussion and contention than the sealing operation and methods of assessing the efficacy of sealing. It is not surprising therefore that the test methods vary widely in principle and the greatest caution must be exercised in the interpretation of test results.

Dye spot testing

This was the earliest form of sealing test and was originally devised to detect the presence of an unsealed coating produced by the chromic acid process. The dye was applied by a wetted "indelible pencil" (containing Methyl Violet dye). The marked surface was then wiped with a damp cloth. In the presence of an unsealed coating the violet stain could not be removed. When the chemical sealing techniques were introduced in the early 1930s the same test was employed—but in this case the dye mark was completely removable by washing if the sealing had been properly carried out. Improved sensitivity was later achieved using a solution of Anthraquinone Violet in chloroform, and this was in turn succeeded by the Scott Test in which the surface to be tested was first spotted with an acid fluoride solution, washed and then spotted with a red dye Aluminium Fast Red B3LW, and again washed.

The principle of pre-spotting with acid has recently achieved international recognition and the following reagents have been specified:

(a) Sulphuric acid ($d = 1.84$)   25 ml/l
       Potassium fluoride   10 g/l
or (b) Hydrofluosilicic acid ($d = 1.29$ g/ml)   25 ml/l

After 1 minute the spot is washed off and the surface again dried.

A spot of one of the following two dyestuff solutions is then placed on the area that was previously acid spotted.

(a) Aluminium Blue 2LW pH 5.0 ± 0.5 5 g/l
or (b) Aluminium Red B3LW pH 5.7 ± 0.5 10 g/l

After 1 minute the dye spot is washed off and the surface rubbed with water and a mild abrasive (magnesia or whiting) for 20 seconds. After further rinsing and drying the spot is examined and the intensity of the stain (if any) is compared with a standard colour chart. A colour intensity of 2 or less is considered satisfactory. This test may give misleading results with some anodic coatings of less than 3 μm thickness due to coating dissolution by the acid-spotting solution.

MEASUREMENT OF ADMITTANCE OR IMPEDANCE

It has already been mentioned that the anodic oxide coating is an electrical insulator. In the unsealed condition the pores to the base are open and, if wetted, will pass a current without difficulty. During the hydration sealing process the pores are closed and the electrical resistance increases. Instruments are available for measuring either the admittance (i.e. the electrical conductivity for AC) or the impedance (i.e. the electrical resistance to the passage of AC).

The impedance test (in North America) and the admittance test in most other countries are widely used for production control purposes.

Taking the admittance test as an example, the conductivity of the coating will depend on two factors:

1. The inherent specific conductivity of the coating when wetted by the test solution.
2. The thickness of the coating. Thicker coatings offer greater resistance, i.e. they reduce the conductivity of the system. It is therefore necessary to check the coating thickness at a given point before setting up the admittance test at the same point.

The test equipment consists of a probe and a pointed screw clamp which are connected to the test instrument which applies AC at 1 kHz. An adhesive rubber ring is applied to the selected area of the coating and the cell thus formed is filled with a solution containing 35 g/l potassium sulphate. The screw clamp is used to make a connection to the basis metal of the test piece by piercing the coating. The probe is put into the cell and the reading of admittance in microsiemens is read off the instrument dial. The test is carried out at room temperature and this should be noted so that a temperature correction can be made. The instrument reading is based on the use of a standard rubber ring having an enclosed area of 133 mm^2. The instrument suppliers provide full instructions for calibrating the instrument and calculating any correction factors.

At one time the validity of this test was widely accepted, but more recently some caution is being exercised in interpreting the results obtained. An admittance not exceeding

$$\frac{500\,\mu\text{sec}}{T}$$

where T is the thickness of the coating is considered acceptable but the addition of bloom-preventing agents to sealing solutions and the adverse effects of phosphate and silicate in hot-water sealing are not always reflected by a high admittance value. As a result it is possible to obtain good results by the admittance test on unsatisfactorily sealed coatings.

As the sealing conditions, quality of sealing water, etc., vary from plant to plant it is recommended that a sealing procedure that gives good results when tested by one of the acid-dissolution methods should first be established. These satisfactory coatings shall then be submitted to the admittance test and the "pass" figure recorded. This figure can be used for production-control purposes but the coatings should also be checked periodically to ensure that they still pass the acid-dissolution test.

It must be mentioned that the admittance value on a given coating tends to decrease with time due to the so-called ageing (a continuation of the hydration sealing process that occurs at room temperature and in a moisture-containing atmosphere). After about 2 months an originally unsealed coating will have about the same admittance as a sealed coating.

Sulphur-dioxide/humidity test

The deterioration of anodized aluminium in an industrial atmosphere is mainly due to chemical attack by sulphurous and sulphuric acid dissolved in the dew that forms almost daily in the U.K.

Although the operation of the Clean Air Act has greatly reduced the amount of solids in the atmosphere derived from burning solid fuel, the sulphur dioxide level is still maintained by the burning of oil fuel.

The introduction of a sulphur dioxide/humidity test was a logical step in the development of accelerated tests for the quality of anodic coatings. Cabinets can be purchased for this test which is carried out at $25°C \pm 2°$ at a relative humidity of less than 100% and not less than 95% with a gas content of 0.5-2.0% by volume sulphur dioxide. The usual exposure time is 24 hours, after which the test sample is examined visually. Any thin superficial bloom is first removed by gentle wiping with a soft damp cloth. If, after this, the surface has a milky white appearance it is evident that some permanent deterioration has occurred. The permissible degree of attack, if any, is usually agreed between the anodizer and the purchaser.

Acidified sulphite test (Kape test)

The 24-hour period needed for the sulphur dioxide/humidity test led to investigations into the possibility of accelerating the process. This was achieved by using an acidified solution of sodium sulphite made up as follows:

To a 10-g/l solution of anhydrous sodium sulphite in distilled or purified water, an addition of glacial acetic acid (20 ml/l to 40 ml/l) was made to bring the pH to 3.6-3.8. A 2.5 M solution of sulphuric acid (10 ml/l to 15 ml/l) was then added to reduce the pH to 2.5 at room temperature.

The sample to be tested was immersed in the solution at 90-92°C for 20 minutes, after which it was washed, dried and examined visually as for the sulphur dioxide/humidity test.

The above version of the test was found to vary in its effect, and being subjective by nature could cause disagreements on the interpretation of the results.

The test was later modified by subjecting the sample to a pre-dip in a 50% by volume solution of nitric acid at room temperature for 10 minutes. This sensitized the coating and gave better discrimination between various degrees of sealing. Still more important, the test could be used quantitatively and is so used today. The specimen (of measured area) is weighed before and after the acid sulphite attack. A weight loss not exceeding 20 mg/dm^2 is considered satisfactory.

Acetic acid/sodium acetate test

On the continent of Europe a different type of acid-attack solution was developed using a solution containing

Glacial acetic acid	100 ml
Sodium acetate	0.5 g
Water	to 1 litre
pH 2.2-2.3	
Temperature	Boiling

The standard time of immersion is 15 minutes and the weight loss per unit area is determined. A figure not exceeding 20 mg/dm^2 is satisfactory. This test was used for many years without a pre-dip in nitric acid but it has now been agreed at international level that the pre-dip shall be included in the test.

Chrome/phosphoric acid test

Both of the acid-attack tests previously mentioned have the disadvantage that they must be carried out in a fume cupboard due to the evolution of sulphur dioxide and acetic acid vapours respectively.

Both tests are tending to be replaced by the chrome phosphoric acid test which was developed in the U.S.A. and is based on the use of the chrome/phosphoric acid mixture used to determine coating mass (see page 103). For this test the solution is distilled or purified water is made up as follows:

Phosphoric acid ($d = 1.75$)	35 ml/l
Chromic acid CrO_3 (Analytical reagent quality)	20 g/l
Temperature	37°C

The time of immersion is 15 minutes. The weight loss should not exceed 30 mg/dm^2.

A nitric acid pre-dip is not required in this test but nevertheless a note on the subject of this pre-dip will not be out of place.

It is known that the pre-dip in nitric acid results in some weight loss of the test specimen and it is worthwhile determining this loss in addition to the acid-attack weight loss. The figure rarely exceeds 10 mg/dm^2. A higher figure may indicate that there is an excess of soluble bloom on the coating surface, which may be due to a softer than usual coating. Such coatings can "chalk" on exposure to the atmosphere. A high nitric acid pre-dip weight loss must therefore be treated with suspicion. Coatings that have been sealed and submitted to one of the sealing smut-dissolution processes will, of course, show a very small weight loss in the nitric acid pre-dip but even in this case, if the sealing is defective it will be detected by an excessive weight loss in the acid-attack solutions. It is not surprising therefore that the acid-attack methods have been adopted as the referee test for satisfactory sealing.

CORROSION RESISTANCE

It has always been difficult to devise accelerated corrosion tests that can be correlated with long-term exposure. Tests for anodic oxide coatings are no exception, but two methods have been adopted and are mostly used in the automobile industry.

Acetic acid salt-spray test

This is particularly suitable for coatings up to 5 μm in thickness and involves exposure in a suitable cabinet to a spray-mist derived from a solution containing 50 ± 5 g sodium chloride with the pH adjusted to 3.2 ± 0.1 by the addition of glacial acetic acid. The cabinet atmosphere is maintained at 35°C and an exposure time of 24 hours is usually specified. No pitting of the coating should be visible after spraying (see Table 10).

TABLE 10* EFFECT OF FILM THICKNESS ON RESISTANCE TO ACETIC ACID-SALT-H_2O_2 SPRAY TEST OF ANODIZED 99.5% ALUMINIUM AND 99.99% ALUMINIUM-1½% MAGNESIUM

99.5% aluminium		Aluminium-1½% Mg based on 99.99% aluminium	
Film thickness (μm)	Time required to produce pitting (hr)	Film thickness (μm)	Time required to produce pitting (hr)
5	20	5	95
8	60	8	185
14	205	10	235
19	480		
26	Over 1300		

*Brace, A. W. and Pocock, K., "Methods of testing anodic coatings on aluminium", *Trans. Inst. Met. Finishing*, **35**, 277-94 (1958).

CASS test (Copper-accelerated acetic acid salt-spray test)

The acetic acid salt spray is a relatively mild reagent. Where thicker films require testing a more active reagent for the spray solution is made up as follows:

Sodium chloride	50 ± 5 g
Cupric chloride	0.26 ± 0.02 g
Water (preferably purified)	to 1 litre

pH 3.2 ± 0.1 adjusted by the addition of glacial acetic acid

The cabinet temperature is $50 \pm 1°C$ and the standard exposure time is 8 hours. A rating system (BS 3745) has been devised to enable a performance number to be allocated to corroded surfaces. Ratings of at least 8 for 15-25 μm coatings and at least 6 for 5-15 μm coatings are acceptable.

Suggested designs and the standardized operating conditions appear in the appropriate British and International Standards (see Appendix IV).

Salt spray tests

Some use is still made of testing in spray from a 5% or 10% solution of sodium chloride. It may take several thousands of hours before signs of pitting become apparent on well-produced films so that the test is unsuitable for production control. By comparison, a good standard of nickel plus chromium plate on a steel basis metal should withstand 90 hours' salt spray.

On some very critical applications of anodized aluminium, for example on stressed components in aircraft, the salt-spray test is followed by fatigue tests to ascertain the effect of any corrosion on the mechanical properties of the coated component.

REFLECTIVITY

The visual appearance of an anodic oxide coating depends upon three important optical properties:

1. The total reflectivity.
2. The diffused reflectivity.
3. The specular reflectivity.

All of these properties can be measured but may require very sophisticated and expensive apparatus. The methods that follow represent a compromise between a high-degree accuracy and commercial production requirements.

Total reflectivity

This is a measure of all the light reflected in all directions by a surface. Measurement within a 1% degree of accuracy can be made with the PRS head (Figure 21). The photo-cell terminals are connected to a galvanometer. The light from the lamp falls on the flat test specimen on which the PRS head is placed and the reflected light activates the photoelectric cell which produces a galvanometer deflection. The head is calibrated on a magnesium carbonate block with a nominal total reflectivity of 100%.

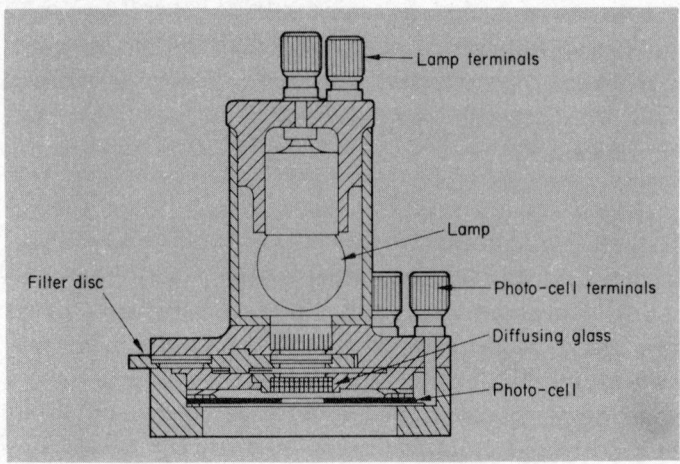

FIGURE 21. PRS HEAD.

Specular reflectivity

On surfaces other than a perfect mirror the light is reflected partly in a specular manner and partly diffused. Very bright surfaces have a small diffused component, whilst etched matt surfaces produce a large diffused and a small specular component. In the latter case the specular component may only be significant when viewed at an acute (glancing) angle to the coating. It follows therefore that to define a method for measuring specular reflection it is also necessary to agree an angle of incidence for the light and an angular range through which the specular reflected beam will be collected and measured.

Testing Anodized Aluminium 115

A range of incident angles has been selected, 20°, 45°, 60° and 85°. The last-mentioned angle is particularly suitable for differentiating very diffuse finishes.

A typical apparatus appears in Figure 15. The surface to be tested, which must be flat, is placed under an aperture in the base of the instrument. The amount of light reflected is recorded on the meter and calibration is carried out on a polished black glass plate.

An alternative instrument with a fixed angle of 45° is based on a modification of the gloss head to DEF 1053. In this case specular reflectivity is compared at an angle of incidence of 45° and a solid angle of acceptance of 0.00125 steradian. The intensity of reflected light, suitably directed, is measured by means of a photoelectric cell in conjunction with a galvanometer equipped with a variable shunt. The reflectivity of a sample, which *must be flat* and at least 15 × 7.5 cm, is measured by placing the gloss head with its base firmly in contact with the sample and noting the reading of the galvanometer. A suitable surface for reference purposes is the hypotenuse of a 45° right angle prism, with total internal reflection of the incident light. With a 2.54 × 2.54 × 3.55 cm prism the absolute specular reflectivity is almost exactly 90%.

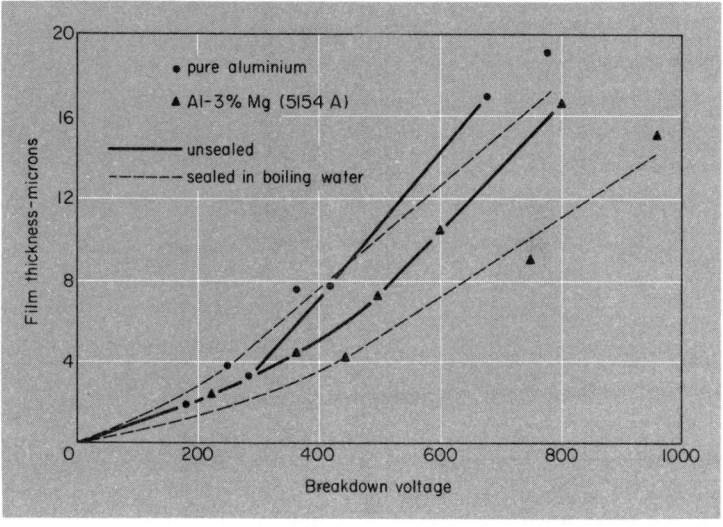

FIGURE 22. BREAKDOWN VOLTAGE–FILM THICKNESS RELATION FOR FILMS PRODUCED UNDER STATIC CONDITIONS; 15 VOL % SULPHURIC ACID, 20°C, 1.86 AMP/FT^2, DRIED IN CABINET AT 110°C.

Diffuse reflectivity

The total reflectivity $T = S + D$ where S is the specular reflectivity and D is the diffuse reflectivity. D is therefore calculated as $T-S$ both of which figures are obtained by the preceding two methods. T, S and D are usually denoted as percentages.

Directional effects

Wrought aluminium products have directional grain or lines which, even on a nominally flat surface, will provide varying reflection factors depending upon the plane of the incident light compared with the directional characteristics of the surface. It is customary therefore to define the direction of incidence, i.e. along the "grain" or across the "grain".

IMAGE CLARITY

The image clarity of anodized aluminium is of major practical importance and can vary markedly on surfaces with the same

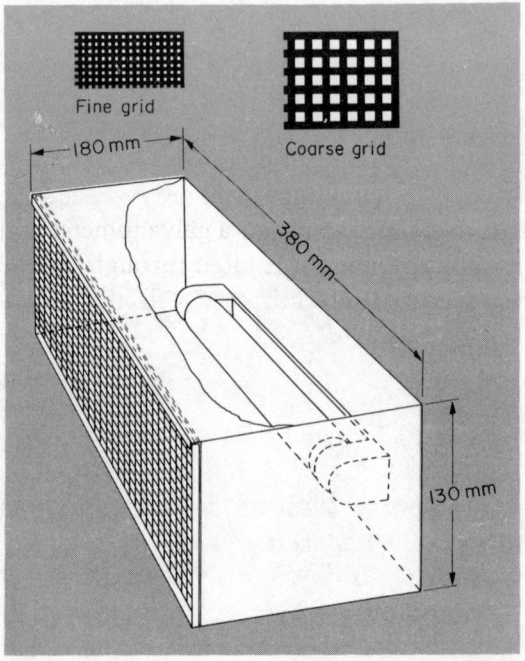

FIGURE 23. GARDAM GRID. (Courtesy: British Standards Institution.)

specular reflectivity. It is conveniently measured qualitatively by the use of a Gardam grid, which test is suitable for routine inspection purposes.

The Gardam grid consists of an oblong box containing a strip-light and open at one side. This side is covered with a glass screen suitably marked with black stripes (e.g. Indian Ink) as shown in Figure 23 and illuminated by the strip-light through frosted or pearl glass. The surface to be measured is usually held by the operator in his hand while he walks away in line with the grid; alternatively the surface may be placed directly under the grid. The operator moves away until he can no longer clearly resolve the grid lines as reflected in the specimen. (The operator must have normal eyesight, i.e. his near point should be 25 ± 2 cm.) The distance from the operator to the grid is then a measure of the image clarity of the specimen. Two grids have been suggested. A coarse grid with ⅜-in. squares and strips and a fine grid with ⅛-in. squares and strips. The latter is preferable for specimens of good image clarity, the coarse grid being for those surfaces having poor image clarity.

INFRA-RED REFLECTIVITY

A method devised by the National Physical Laboratory is fully described in BS 1615 and requires the use of a flat sheet specimen. A thermocouple is attached to the face opposite that to be tested and is then painted with matt black paint of known infra-red absorptive capacity (usually about 0.96). The side to be tested is subjected to intermittent heating for fixed times and intervals and the temperature rise is recorded from the readings of a galvanometer connected to the thermocouple. The specimen is rotated through 180° and the radiation exposure is repeated. The infra-red reflectivity is calculated from the formula

$$1 - a \frac{Dt}{Df}$$

where a is the absorption factor of the black-painted side,
Dt is the mean difference between the temperatures recorded on opening and shutting off the radiation on the test side,
Df is the temperature difference recorded on the black side.

The requirements for infra-red reflection usually include a note of the temperature of the radiation source, and such a source must be used for the above test.

LIGHT FASTNESS TESTS

Experience has shown that the fastness to light of a given coloured sample of anodized aluminium will vary in different parts of the world depending on the intensity and duration of sunlight, the temperature rise of the specimen and, to complicate the matter, the degree of chemical attack on the dyestuff and/or on the coating.

Accelerated tests of fastness using intense sources of irradiation are not entirely satisfactory but they serve to eliminate unsatisfactory colours. However, colours that pass the requirements of these tests must eventually have their true performance verified by long term exposure.

The principles of accelerated light fast testing vary only in the source of irradiation which may be a xenon lamp, mercury-in-quartz arc, or carbon arc, all of which have different spectra. The pieces to be tested are half masked and, in the case of the xenon lamp and carbon arc, a set of half-masked standard blue colours (BS 1006) is placed at the same distance from the light source. Exposure is continued until the degree of colour change of the test specimen or the no. 6 cloth (whichever occurs first) is equivalent to Grade 3 on the Grey Scale of BS 2662c. If the no. 6 cloth changes first then a fresh half-masked no. 6 cloth is used and so on until the specimen also exhibits a colour change equal to Grey Scale Grade 3. The number of no. 6 cloths used is noted and the following light fastness ratings are allocated:

Number of no. 6 cloths faded	Light fastness number of sample
1	6
2	7
4	8
8	9
16	10

If the sample fades before the first no. 6 cloth then the lower number cloths are examined and the cloth showing the Grade 3 fading is allocated as the light fastness number. For prolonged outdoor exposure (10 years minimum) a light fastness number of 9 or more is essential. For indoor use a figure of 5 is acceptable.

In the case of the mercury arc in quartz (the so-called "UVIARC" test) the intensity of ultra-violet radiation is very high and causes disintegration of the blue test cloths. It is therefore necessary to use a standard specimen of known light fastness (usually prepared by electrolytic colouring or integral colour anodizing) for control purposes. Only the best colours will survive 24 hours' exposure to this lamp so that the test can be used for production-control purposes.

ABRASION RESISTANCE

This is an important quality of anodized aluminium but the standardization of test methods has been fraught with difficulty due to lack of comparability between the results obtained with apparently identical pieces of equipment. Recent developments suggest that a standard abrasion resistance test piece will be established for calibrating the performance of test apparatus.

Two methods of test are gaining favour, one based on the penetration of the coating by a jet of abrasive powder under controlled conditions and the other on the rate of removal of the coating by an oscillating abrasive coated band, again under carefully standardized conditions.

Abrasive jet test (Modified Schuh and Kern Method)

A version of this test appears in BS 1615 where the apparatus and method are described in detail (see Figure 24). In brief, the coating which need not be on a flat surface is subjected to a stream of dry air carrying dry silicon carbide powder of 106-μm grade, at a flow rate of 40 l/min. The angle of impingement is 45° and the amount of grit in the air stream is adjusted to 25 g/min.

When the coating becomes penetrated to the basis metal a grey spot appears in the abraded area and the abrasive jet is then shut off. The weight of the silicon carbide grit used is determined so that for a measured coating thickness a figure of grams of abrasive per micron can be calculated. Some typical results are shown in Table 11.

Variations in jet design and actual wear of the jet in use will alter the performance of the equipment. It is for this reason that "standard" test pieces for calibration are required.

For thick hard anodized coatings an air flow rate of 100 l/min is used so as to avoid an unduly long time for penetration.

Abrasive-wheel test

This is a more recent development than the abrasive-jet method. It requires a flat test piece which is abraded by the oscillation of an abrasive wheel. The coating is gradually worn away and it is practicable and sometimes desirable to determine the thickness of coating removed after a given number of strokes, thus providing a profile of the variation of abrasion resistance throughout the coating.

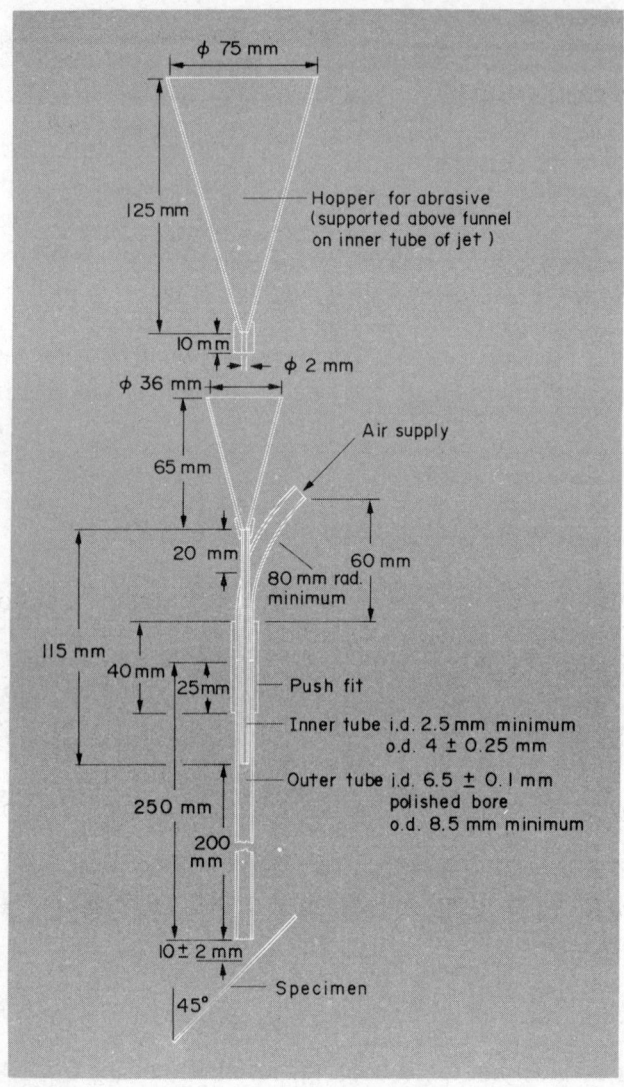

FIGURE 24. ABRASION RESISTANCE ASSESSMENT BY THE MODIFIED SCHUH AND KERN METHOD. (Courtesy: British Standards Institution.)

Alternatively, the number of strokes required to penetrate the coating to the basis metal can be recorded or the thickness of coating removed by an agreed number of strokes can form the basis of acceptance tests. Instead of measuring the loss of coating thickness,

TABLE 11 EFFECT OF ANODIZING CONDITIONS ON SCHUH AND KERN*
ABRASION RESISTANCE OF ANODIC COATINGS

Anodizing conditions	Abrasion resistance (g)		Film thickness (μm)	Specific abrasion resistance (g/μm)
	Values	Average		
Chromic acid (Bengough-Stuart) (DEF 150)	172, 157, 169	166	5	33
20 min in 3.3 N H_2SO_4, 21°C (70°F), 1.5 amp/dm²	376, 406, 382	388	10	39
20 min in 7.5 N H_2SO_4, 21°C (70°F), 1.5 amp/dm²	200, 193, 206	200	10	20
20 min in 3.3 N H_2SO_4, 15.5°C (60°F), 1.5 amp/dm²	579, 536, 574	563	10	56

*Brace, A. W. and Pocock, K., "Methods of testing anodic coatings on aluminium", *Trans. Inst. Met. Finishing*, **35**, 277-94 (1958).

the weight loss of the specimen can be determined although this alternative is perhaps better suited to laboratory investigations.

The pressure of the wheel on the coating can be adjusted and in order to present a fresh abrasive wheel surface after each double stroke the wheel indexes automatically. The circumference of the wheel is sufficient to allow for 400 strokes before the abrasive band covering of the wheel has to be replaced. A typical apparatus is shown in Figure 25.

TESTING THE CONTINUITY OF ANODIC OXIDE COATINGS

Film continuity methods are primarily used for the evaluation and control of continuously anodized strip, foil and wire.

A chemical method of detecting breaks in the anodic coating involves immersion in

Crystalline copper sulphate	20 g	
Hydrochloric acid	20 ml	5 minutes at 15-20°C
Water to 1 litre		(59-68°F)

Increasing the concentration of each chemical to 10% can reduce the time required to 15 seconds. The reagent does not affect an anodic film; black spots appear where no anodic film is present.

This method will also detect cracks in the coating caused by overheating or bending.

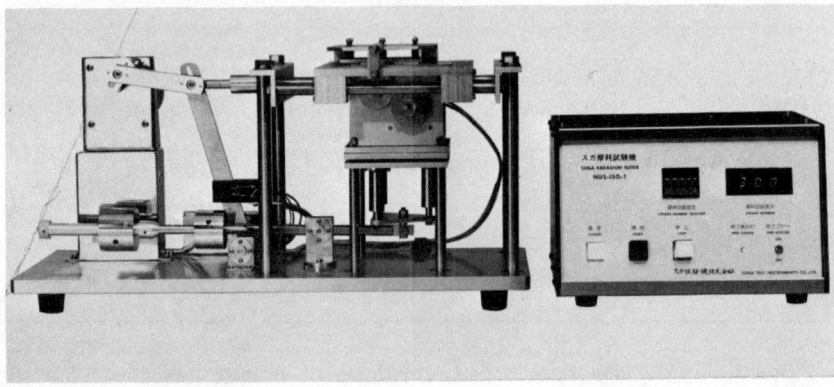

FIGURE 25. ABRASIVE WHEEL APPARATUS FOR ANODIZED ALUMINIUM. TOP LEFT IS THE RECIPROCATING MECHANISM FOR THE CARRIAGE (TOP RIGHT) THAT CARRIES THE SPECIMEN. A DIGITAL READ-OUT DISPLAYS THE NUMBER OF STROKES. (Courtesy: Suga Test Instruments Co. Ltd., Tokyo.)

MEASUREMENT OF RESISTANCE TO CRAZING

For certain special applications where the anodic coating has to be formed it is necessary to control and test the ductility of the coating. The simplest test involves bending the sample over a mandrel of agreed diameter and then visually examining the coating for cracks or crazing. In case of doubt the bent coating can be checked for continuity by the copper sulphate solution described in the previous section.

More useful information on the formability of the coating can be obtained by using a spiral mandrel which appears in an International Standard (see Appendix IV). With this former the minimum radius at which cracking is first visible can be recorded.

ELECTRICAL BREAKDOWN VOLTAGE

In the standard test given in BS 1615, a hard, spherical electrode of $\frac{1}{16}$ in. radius with a 50-75-g load is in contact with an undeformed anodized test piece. Positive electrical contact is maintained between the secondary winding of a transformer and (a) the electrode, (b) the basis metal on the underside of the test piece. Voltage is increased uniformly at a rate not exceeding 25 V/sec until electrical breakdown occurs.

For wires, the usual test method (given in BS 1615, Appendix U) involves making a standard twist joint with two wires and gradually increasing their potential difference until breakdown occurs.

The breakdown voltage of unsealed films varies with humidity, so that humidity control is required.

Chapter 14

The Properties of Anodized Aluminium

The surface coating on anodized aluminium consists of alumina (aluminium oxide) with modifications derived from the electrolyte or from the alloying elements of the basis metal. The coatings formed in sulphuric acid contain 13-15% sulphate, those in phosphoric acid about 6% phosphate, but those in chromic acid only about 0.2% chromate. Some water is chemically retained in the coating at the sealed surface, which corresponds mainly to the formula $Al_2O_3 \cdot H_2O$. The unsealed coating is virtually anhydrous.

THICKNESS

The anodic coating always includes a thin, hard, compact "barrier" layer close to the metal. In non-solvent electrolytes this is the only coating present and has a thickness in Ångstrom units of about 14 times the anodizing voltage. In solvent electrolytes (i.e. those in which alumina is slightly soluble) an outer, much thicker, layer forms (Figure 26). Normally the maximum thickness is about 0.0015 in., but special techniques enable coatings up to about 0.010 in. to be formed. BS 1615 specifies grades of anodizing according to coating thickness. Not all combinations of alloys and anodizing processes will give the thicker coating. Sulphuric acid anodizing with straightforward techniques will give 25-μm coatings or thicker on all alloys except the aluminium-copper alloys. Chromic acid anodizing by the standard processes will not give greater than 5 μm with pure aluminium.

DENSITY

The thin barrier layer has a density of 3.2 g/cm^3. For many solvent-type electrolytes with 10-20% pore volume the true density

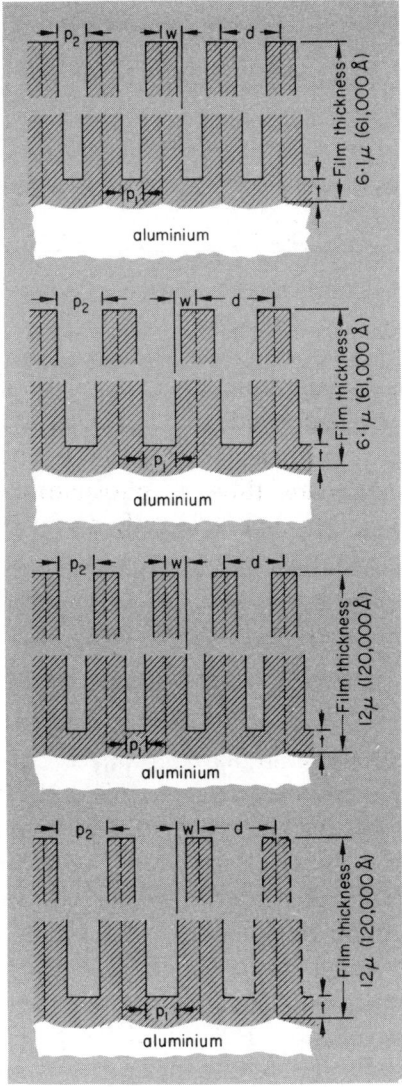

A. Thin Film *Quantity of current:* 20 A min/dm^2

Current density: 1 A/dm^2
Anodizing time: 20 min
Voltage: 13 V
$t = 130\text{--}180$ Å
$w = 105$ Å
$p_1 = 100$ Å
$p_2 = 100 + x$ Å (x dependent on thickness of film)
$d = 310$ Å

Current density: 2.5 A/dm^2
Anodizing time: 8 min
Voltage: 17 V
$t = 170\text{--}240$ Å
$w = 135$ Å
$p_1 = 250$ Å
$p_2 = 250 + x'$ Å ($x' > x$ dependent on thickness of film)
$d = 520$ Å

B. Thick Film *Quantity of current:* 40 A min/dm^2

Current density: 1 A/dm^2
Anodizing time: 40 min
Voltage: 13 V
$t = 130\text{--}180$ Å
$w = 105$ Å
$p_1 = 100$ Å
$p_2 = 100 + y$ Å ($y > x$ dependent on thickness of film)

Current density: 2.5 A/dm^2
Anodizing time: 16 min
Voltage: 17 V
$t = 170\text{--}240$ Å
$w = 135$ Å
$p_1 = 250$ Å
$p_2 = 25 + y'$ Å ($y' > y$ dependent on thickness of film)

FIGURE 26. THE STRUCTURE OF DC ANODIC FILM. INFLUENCE OF CURRENT DENSITY, VOLTAGE AND FILM THICKNESS ON PORE FORMATION, CELL SIZE, AND THICKNESS OF BARRIER LAYER AT OTHERWISE CONSTANT OPERATING CONDITIONS. ELECTROLYTE: 250 g/l H$_2$SO$_4$, 0.13 g/l DISSOLVED ALUMINIUM AT 17°C, AGITATED BY AIR. Lenz, D., *Aluminium*, 1956, **37** (3), 126-35; (4), 190-201.

(determined by Archimedes' principle) for unsealed coatings is fairly constant (2.9-3.0). The apparent density of the sealed coating varies more widely; Table 12 shows the variation with thickness for a typical set of anodizing conditions, but the aluminium-copper alloys may give values under 2.0 with nearly 50% pore volume. High anodizing temperatures and high acid concentrations (i.e. high coating solu-

TABLE 12 APPARENT DENSITY OF SEALED ANODIC COATINGS
(30-MINUTE ANODIZED PANELS)

Sulphuric acid electrolyte		Current density (amp/ft^2)	Film thickness (μm)	Apparent density
Temperature (°C)	Conc. (% by wt.)			
10	15	15.8	18.1	2.54
	30		16.8	2.53
	50		19.0	2.39
21	5	15.8	16.3	2.66
	15		16.5	2.64
	30		17.1	2.36
	50		18.7	2.08
30	5	15.8	15.6	2.68
	15		15.0	2.53
	30		15.6	2.04
	50		10.0	1.93
50	5	15.8	9.7	1.9
	15		3.4	1.8
			1.5	1.5
21	15	4.6	4.1	2.8
		15.8	14.0	2.68
		33.5	29.7	2.66
		64.2	66.0	2.46
30	15	15.8	14.9	2.50
		33.5	30.5	2.41
		64.2	58.5	2.18
50	15	15.8	2.5	2.1
		33.5	6.5	2.1
		64.2	12.3	2.1

Spooner, R. C., "The anodic treatment of aluminium in sulfuric acid solutions", *J. Electrochem. Soc.* **102** (4), 156-62 (1955). Reprinted by permission of the publisher — The Electrochemical Society Inc.

The Properties of Anodized Aluminium

bility) give low densities, due to a high air-filled pore volume. With sulphuric acid electrolytes containing chlorides, under continuous anodizing conditions, values of 1.7-1.95 have been obtained on aluminium. Some commercial specifications stipulate a minimum coating density. Appropriate values for typical resistant coatings are given in Table 12. The porosity referred to above is the microporosity inherent in the structure of the anodic film formed in solvent electrolytes, and is not to be confused with that harmful macroporosity — rarely encountered — due to metal defects or extreme operating conditions.

The general effects on the properties of anodic films caused by varying the bath conditions are shown in Table 13 for solvent electrolytes.

TABLE 13 EFFECT OF CHANGING BATH CONDITIONS ON PROPERTIES OF ANODIC FILMS

Change of* bath conditions	Effect on		
	Softness,† porosity, elasticity, and absorptive characteristics	Protective power	Solubility rate of metal in bath
Rise of temperature of the bath	Increase	Decrease	Rapid increase
Increase in acidity of the solution	Increase	Decrease	Increase
Increase of current density (for the same duration of the process)	Change depends on temperature and agitation	Increase	Decrease if agitation is adequate
Increase of duration of the treatment (at the same current density)	Slight increase	Increase	Increase
Alternating current in place of direct current	Increase	Decrease	Increase

*It is assumed that with alteration of one of the three variables (temperature, concentration and current density) the other two are kept constant by altering other operating conditions, e.g. voltage applied, agitation.

†The decrease of the softness, porosity and elasticity of the anodic film corresponds, generally, to the increase of hardness, density and brittleness.

Note: The occurrence of a surface bloom indicates poor anodizing practice, such as too high a temperature, too long immersion or insufficient agitation. This bloom in some forms may appear only after some months' outdoor exposure.

MECHANICAL PROPERTIES

Strength

The tensile strength and elongation of the basis metal are not altered by the anodic treatment; with very thin material allowance must of course be made for the thickness of metal converted to oxide and for notch effects. The fatigue strength is reduced by anodizing: with hard anodized aluminium the reduction can amount to as much as 50%, but this may be alleviated, albeit with some loss of hardness, by boiling for 15 minutes in 5% potassium dichromate solution. With ordinary anodizing up to 10 µm, fatigue strength is only slightly reduced even at relatively high stresses, while at low stresses apparent gains have been noted — probably in the corrosion fatigue properties, due to enhanced corrosion resistance.

The alumina film has a significant strength when detached from the metal; it has been used for pressure discs by dissolving the backing metal over a circular area in the middle of a plate (see page 98). For barrier-layer films the breaking pressure P_{max} in mmHg has been empirically determined* as

$$P_{max} = \frac{400}{d} \log_{10} \frac{V}{24}$$

where V = anodizing voltage (between 50 and 100 V),
d = diameter of circle (mm).

Flexibility and Hardness

Anodic films cannot be permanently deformed without damage, but if fine crazing is not apparent the coating is said to be flexible. The flexibility and hardness of anodic films are not usually measured directly; both depend to a considerable extent on the anodizing process and conditions, and flexibility usually decreases with increasing hardness. In general, flexibility increases with electrolyte temperature, acid concentration and activity; films produced using alternating current are more flexible than those produced using direct current. Chromic acid films are relatively flexible, as shown by the comparison with sulphuric acid films in Table 14.

The technique for continuous anodizing also makes for flexibility when the films are thin. Substantial forming operations can therefore

*Hauser, U. and Kerner, W., "Easily mounted aluminium oxide foils for windows and backings", *Rev. Sci. Inst.* **29** (5), 380-2 (1958).

TABLE 14* BEND TESTS ON SLC-½H, WITH CHROMIC AND SULPHURIC ACID ANODIZED COATINGS

Electrolyte	Film thickness (μm)	1½ in. dia. mandrel		1 in. dia. mandrel		½ in. dia. mandrel	
		Compression	Expansion	Compression	Expansion	Compression	Expansion
10% CrO_3 at 30 V and 54°C (129°F)	6	A	A	B	A	C	B
	8	A	A	B	A	B	B
150 g/l H_2SO_4 at 15°C (59°F) and 15 amp/ft²	6	B	C	B	C	C	D
	10	B	C	B	C	C	D

A = No visible damage   C = Crazing
B = Very slight crazing   D = Severe crazing

*Brace, A. W. and Peek, R., "Production and properties of opaque coatings by chromic acid anodizing", *Trans. Inst. Met. Finishing,* **34**, 232-52 (1957).

be achieved with rounded tools on the anodized aluminium without easily visible crazing—bottle tops and trim for radio sets are examples. In most applications fine crazing is not detrimental, as it only slightly reduces corrosion resistance.

Even under the most favourable conditions for producing flexible coatings the maximum elongation before cracking occurs does not exceed 0.3%.

The hardness is 7-9 on Moh's scale, the anodized surface marking glass and steel; the hardest films of all mark chromium plate.

Measurements of hardness by the use of diamond pyramid micro-indentations (taken on a cross-section) give a wide range of results, even with sulphuric acid anodizing. The values usually lie between 100 and 400 VPN, with hard anodizing giving values up to about 500 VPN.

Crazing of anodic films is likely to occur at sharp corners and edges. This can materially affect the local durability of surfaces exposed to a corrosive environment.

The flexibility of hard anodized coating is extremely low and there may even be audible fine crazing, particularly at corners, when the component is withdrawn from the cold anodizing bath and allowed to warm up at room temperature.

Abrasion Resistance

Abrasion resistance usually increases with the hardness of the film and with the film thickness, the outer layers of any film being the softer and less abrasion resistant. For comparative purposes a specific abrasion resistance is commonly used, i.e.

$$\frac{\text{Abrasion resistance of complete thickness of film}}{\text{Film thickness}}$$

The test methods used to date do not permit reliable comparison between the results obtained on different pieces of apparatus, even if they are nominally of the same type. No attempt is made, therefore, to give typical values of specific abrasion resistance for different alloys. The test method given in BS 1615 (in which abrasive particles are projected by an air blast on to the metal surface under controlled conditions) shows differences between coatings sealed by different methods and shows that unsealed coatings have greater abrasion resistance than sealed coatings. Table 15 compares the abrasion resistance of sulphuric acid films with those of chromic acid films, utilizing a blast method.

TABLE 15 EFFECT OF ANODIZING CONDITIONS ON SCHUH AND KERN*
ABRASION RESISTANCE OF ANODIC COATINGS

Anodizing conditions	Abrasion resistance (g)		Film thickness (μm)	Specific abrasion resistance (g/μm)
	Values	Average		
Chromic acid (Bengough-Stuart) (DEF 150)	172, 157, 169	166	5	33
20 min in 3.3 N H_2SO_4, 21°C (70°F), 1.5 amp/dm^2	376, 406, 382	388	10	39
20 min in 7.5 N H_2SO_4, 21°C (70°F), 1.5 amp/dm^2	200, 193, 206	200	10	20
20 min in 3.3 N H_2SO_4, 15.5°C (60°F), 1.5 amp/dm^2	579, 536, 574	563	10	56

*Brace, A. W. and Pocock, K., "Methods of testing anodic coatings on aluminium", *Trans. Inst. Met. Finishing*, **35**, 277-94 (1958).

The specific abrasion resistance increases with voltage with all solvent-type electrolytes. In general, hard anodic coatings have a specific abrasion resistance some 2 to 3 times that of ordinary coatings; the variation between alloys is less marked with the hard coatings. The increased resistance of hard anodized coatings to rubbing abrasion is relatively greater than to blast abrasion.

Friction

The seizing of two aluminium mating surfaces may often be avoided if one surface is anodized (preferably hard anodized). Both surfaces may be anodized when lubrication is available, but it is recommended that coatings of different hardness be used.

OPTICAL PROPERTIES

Three reflection characteristics of anodic oxide coatings are important commercially:

1. The total reflectivity.
2. The specular reflectivity.
3. The diffused reflectivity.

The total reflectivity of bright anodized aluminium on either etched or polished surfaces usually exceeds 80%. From Table 17 it will be seen that this compares favourably with competitive materials such as stainless steel and chromium-plated brass. It is lower than silver-plated brass which, however, readily tarnishes unless it is lacquered. Silvered glass also gives figures similar to silver-plated surfaces, but the fragility of glass can be a disadvantage. The total reflectivity of anodized aluminium decreases with an increase in coating thickness so that it is sometimes necessary to compromise between a thin coating to obtain the highest reflectivity and a thicker film to provide the necessary corrosion resistance.

For the production of mirror-like coatings a high specular reflection factor is essential with the minimum of diffused reflection (see Table 16). From this table it will be seen that for "mirror" finishes the super-purity (99.99% Al) based alloys give the best results. Where a lower mechanical strength can be tolerated the use of 99.99% unalloyed aluminium is preferred. It gives marginally higher reflectivity figures (see Table 16).

TABLE 16* RELECTIVITY OF ANODIZED ALUMINIUM OF
DIFFERENT PURITIES

Film thickness (μm)	Purity (%)											
	99.9				99.8				99.5			
	Specular			Total	Specular			Total	Specular			Total
	B	A	O	T	B	A	O	T	B	A	O	T
2	90	87	88	90	88	68	83	89	75	50	70	86
5	90	87	88	90	88	63	85	88	75	36	64	84
10	90	86	88	89	88	58	85	87	75	26	61	81
15	90	85	88	88	88	53	85	86	75	21	57	77
20	90	84	88	88	88	57	85	84	75	15	53	73

B = Reflectivity of clean surface before anodizing.
A = Anodized surface.
O = Surface after stripping off anodic film if chromic/phosphoric acid mixture.
T = Total reflectivity after anodizing.

*Scott, B. A. and Bigford, H. M., "Bright anodized aluminium surfaces", Paper No. 4, ADA Conference on Anodising, September 1961.

Diffused reflectivity surfaces are used not only for purely optical applications, but also to provide attractive matt finishes for decorative purposes. The original phosphoric/sulphuric acid bright etching solution produces this type of finish.

There are many other decorative finishes where a high specular reflectivity with some diffused component can be tolerated — these are generally classified as "bright" anodized coatings. The finishes on BT2 alloys in Table 17 are typical of this class and they represent the major portion of the work that is subjected to brightening processes. Motor car and consumer-durable bright trim are applications which often pass unrecognized by the general public due to their resemblance to chromium plating.

The effect of alloy composition on the reflection characteristics of various alloys is shown in Table 18.

TABLE 17 TYPICAL REFLECTIVITIES OF SUPER-PURITY AND HIGH-PURITY ALUMINIUM-BASE ALLOYS AND OTHER BRIGHT TRIM MATERIALS

Material	Treatment	Film thickness (µm)	Reflectivities			
			Total	Specular		
			% of incident light	% of incident light	% of total reflectivity	
99.98% Al + 0.9-1.4% Mg (BT2 alloy)	Polished, "Brytal"	4	90	77	85.6	
	Polished, "Brytal"	7-8	90	76	84.4	
	Polished, "Brytal"	10-12	90	75	83.3	
	Polished, "Brytal"	20	90	74	82.2	
Al + 0.8-1.5% Mg with Si + Fe limited to 0.15 each (formerly BT2)	Polished, "Phosbrite" 159	5	88	65	73.9	
	Polished, "Phosbrite" 159	7-8	87	62	71.3	
	Polished, "Phosbrite" 159	10-12	87	60	69.0	
	Polished, "Phosbrite" 159	20	84	52	61.9	
Silver-plated brass	Polished	—	98	86	87.8	
Chromium-plated brass	—	—	65	62	95.4	
Stainless steel	Polished	—	60	53	88.3	

TABLE 18* INFLUENCE OF COMPOSITION OF MATERIAL ON RESPONSE TO BRIGHT ANODIZING WITH "PHOSBRITE 159"

Material	Film thickness (μm)	Total reflectivity (T)† (%)	Specular reflectivity (S)† (%)	Specular ratio (S/T)† (%)
99.99% Al-1¼% Mg	–	87.8	86.5	98.6
	10	82.0	80.5	98.2
	18	80.7	79.2	98.1
1080A	–	89.0	87.8	96.6
	5	83.0	81.8	98.5
	10	82.0	80.2	97.9
1050A (anodic qual.)	–	85.5	82.2	95.1
	5	78.0	68.5	87.9
	10	73.0	59.5	81.5
1200	–	87.5	83.7	95.7
	5	77.5	65.5	84.5
	10	71.7	57.2	79.9
3103	–	84.5	81.0	96.1
	5	72.7	42.5	65.3
	13	62.2	33.5	53.2
5154A	–	85.0	80.0	94.0
	6	76.0	60.5	79.7
	14	71.0	47.5	66.9
6061 type	–	84.7	77.2	91.0
	5	71.0	50.5	71.2
	10	65.0	37.0	56.8

*Brace, A. W., *A.E.S. 46th Annual Tech. Proc.*, p. 216.
†Guild photometer values (integrating sphere apparatus).

REFRACTIVE INDEX

The refractive index of the unsealed film is about 1.59; after hot-water sealing it rises to about 1.62.

FASTNESS TO LIGHT OF COLOURED COATINGS

From the remarks on this subject on page 78 it will be seen that for prolonged outdoor exposure to sunlight and weather where a life of 10 years or more is specified, it is recommended that one of the following colouring processes should be used:

1. Integral colour anodizing.
2. Electrolytic colouring (subject to the recommendations of the process licensors).
3. Ferric ammonium oxalate solution (for a gold colour).
4. Cobalt acetate/potassium permanganate bronze — now rarely offered.

For outdoor exposure up to about 10 years there is a limited range of colours which, if correctly applied to coating of adequate thickness, will provide a useful colour range (see Table 9).

All of the above colouring processes are suitable for indoor use, and indeed the range of applicable organic dyestuffs can be safely extended to include all colours with a light fastness number of 5 or more. For items that have a short service life, or where some colour change is acceptable, the light fastness number is unimportant—a typical example is the range of dyestuffs used to dye aerosol can components.

TABLE 19* HEAT RESISTANCE OF CHROMIC AND SULPHURIC ACID ANODIZED PANELS

Electrolyte	Film thickness (μm)	Temperature (°C)				
		90	130	170	350	580
10% CrO_3, 30 V, 54°C (129°F)	6	A	A	A	A	A
	8	A	A	A	A	A
150 g/l H_2SO_4, 150°C (59°F), 15 amp/ft²	6	A	B	C	C	C
	10	A	B	C	C	C

A = No apparent damage   B = Slight damage   C = Crazing

*Brace, A. W. and Peek, R., *Trans. Inst. Met. Finishing,* **34**, 232-52 (1957).

THERMAL PROPERTIES

Heat Resistance

Anodic films do not blister or peel. At temperatures above about 100°C (212°F) the film may become crazed, because the coefficient of thermal expansion of the coating is only about 20% that of aluminium; also dehydration of coatings sealed by hydration commences above about 400°C; but the melting temperature of alumina is 2050°C (3722°F) — far higher than that of aluminium or the hydrated oxide. This crazing is only objectionable on decorative finishes or where maximum corrosion resistance is needed. Chromic acid coatings craze less visibly than comparable sulphuric acid films, and Table 19 shows that no visible crazing occurs at 580°C (1076°F) on a film 8 μm thick produced by the 10% chromic acid process at 55°C (131°F).

Heat Emissivity and Reflectivity

Aluminium can be used equally well, with difference surface finishes, either to radiate heat or to reflect it.

Emissivity

The infra-red emissivity (heat-radiating ability) of aluminium is only about a tenth of the emissivity of a black body, but can be greatly increased by anodizing. This is because the oxide layer is an effective radiator if it is more than 0.8 μm thick, and its effectiveness increases with increasing thickness.

At 400°C, the emissivity of aluminium with a thick oxide layer ranges from 70% (of black body) at a wavelength of 1.8 μm to less than 30% at λ = 2.5 μm and over 70% at λ = 9 μm. At the temperature of water or steam radiators the emissivity is even greater (approaching 100% for sulphuric acid anodized aluminium) with an oxide layer 2.5 μm thick, at wavelengths greater than 10 μm, so that anodized aluminium is suitable for use in water or steam appliances and heat exchangers.

Reflectivity

When aluminium is used to reflect heat — for example in a domestic radiant electric heater — the anodic oxide layer must be as thin as possible consistent with its ability to protect the metal surface from tarnishing. An oxide layer less than 0.8 μm thick (as mentioned above) is practically transparent to infra-red radiation, and the underlying metal surface, if polished before anodizing, will reflect as much as 95% of the incident radiation (see Figure 27). Even the thin oxide layer, however, will absorb radiation with a wavelength of about 3 μm, and this effect has been attributed to the hydrated surface film on water-sealed oxide layers. For this reason it has been proposed that anodized aluminium heat reflectors should not be sealed in hot water, but should be heated to 160-170°C (320-338°F) for about half an hour after anodizing and waxed while still warm.

The finish given to the metal surface before anodizing also affects reflectivity. The best results are obtained from high-purity material polished and brightened before anodizing.

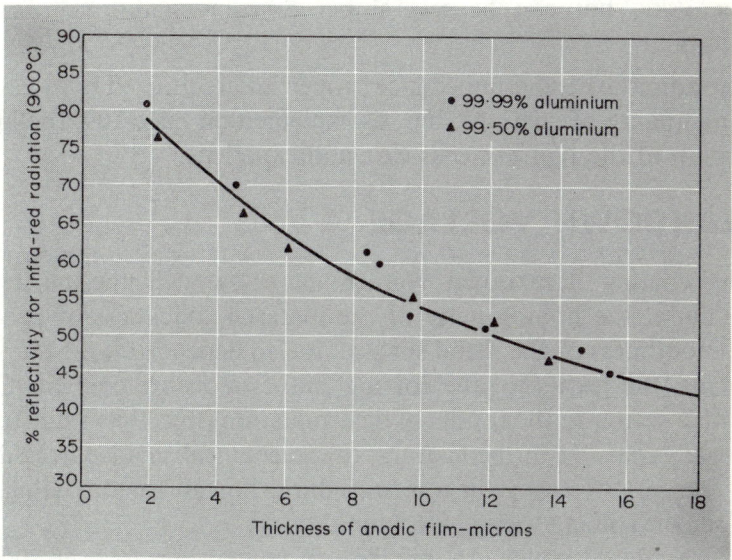

FIGURE 27. REFLECTIVITY OF ALUMINIUM FOR INFRA-RED RADIATION. (SOURCE TEMPERATURE 900°C.)

THERMAL CONDUCTIVITY

Anodic films have about one-tenth the thermal conductivity of aluminium. This is of little significance, due to their thinness, except possibly for hard anodized coatings.

ELECTRICAL PROPERTIES

Unsealed films have a relative permittivity of about 5.0-6.0 for sulphuric acid films and 7.5-8.0 for oxalic acid films, the value rising with temperature and varying with film thickness and anodizing conditions. For barrier-layer films formed in tartaric acid a value of 12 has been found appropriate. The relative permittivity of anodic films is markedly affected by pre-treatment, processing conditions, humidity, etc., and no reliable data can be quoted. The resistance of the porous film decreases with humidity, but this effect is largely eliminated when the film is sealed. The resistivity of the oxide film varies with temperature and is about 10^{13} ohm cm at 250°C (482°F). The resistance may be raised with suitable organic sealing or surface films such as insulating varnishes based on phenolformaldehyde resins, wax, transformer oil, etc.

Power factor losses in anodized aluminium windings are normally about ½%, but can be several per cent in conditions of high humidity.

The admittance and impedance properties are utilized to assess the sealing quality of coatings, but, as explained on page 109, the interpretation of the figures needs care and experience.

BREAKDOWN VOLTAGE

The coating breakdown voltage increases with the purity of aluminium, the homogeneity of the material, thickness of coating and smoothness of the metal surface; it also depends on the alloying elements, the texture of the coating and its moisture content. Considerable scatter in the results of determinations may occur due to (a) intrinsic variations in the coating, (b) mechanical surface irregularities, and (c) differences in test procedure. For unsealed coating the breakdown voltage varies with humidity.

In practice, although AC breakdown voltages up to over 2000 V rms may be obtained on thick lacquer-sealed coatings, most requirements need insulation against only a few volts. Commercial, continuously anodized products with a coating thickness of about 5-8 μm consistently achieve breakdown voltages between turns of over 250 V (i.e. over 125 V for a single thickness of anodizing) and these breakdown voltages are maintained up to over 500°C. Rather higher breakdown voltages are possible on coatings produced under static conditions, as shown in Fig. 22.

Crazing of coatings by bending does not significantly affect the breakdown voltage under reasonably dry conditions.

Precise relationships between film thickness and forming voltage (10^7 V/cm) exist for "barrier" electrolyte coatings (e.g. boric acid or ammonium tartrate), see page 74.

CHEMICAL PROPERTIES

Corrosion Resistance

The corrosion resistance of anodized aluminium is particularly important in the building and shop-front industries, and considerable work has been carried out to ensure that the coating will give satisfactory service. Although anodizers' guarantees against corrosion do not exceed 10 years, experience has proved that useful service of 20-30 years can be expected, provided that accumulated atmospheric deposits are cleaned off at regular intervals—the length of this interval depends upon local conditions, for example a maximum interval of 3 months is recommended for monumental buildings.

To achieve the best corrosion resistance the first requisite is an adequate coating thickness. In the U.K. a minimum thickness of 25 μm is essential in all industrial areas, and as it is often difficult to predict where pre-anodized material is likely to be used it is customary to specify 25-μm coatings for all outdoor work. For some domestic applications, such as double glazing and for shop fronts where washing is frequent, thinner coatings are sometimes used.

In other parts of the world where the atmosphere is less aggressive it suffices to specify 15- or 20-μm coatings.

The efficiency of the sealing of the coating is also important because this process increases the resistance of the coating to chemical attack.

There is another factor rather more difficult to define and test, that is vitally important in this field, namely the "quality" of the coating. During the whole history of anodizing there have been reports of coatings that have become white and chalky after, say, 6 to 12 months' exposure. Even coatings that have been properly tested and passed for sealing quality have failed in this way. There seems to be general agreement among those who have investigated this phenomenon that the quality of the outer layer of the coating has been defective. The most probable cause is the formation of a softer outer coating than normal due to failure to remove the heat evolved at the surface during anodizing—and this in turn is attributed to insufficient and inefficient agitation of the electrolyte.

This inferior type of coating can be detected by one of the abrasion resistance tests which will indicate a lower than usual resistance figure. A chemical spot test is also under investigation.

One of the most corrosion-prone situations is an outdoor surface that is sheltered from the rain so that when specifying the frequency of a cleaning operation it is this situation that should be used as a criterion.

The resistance to a marine environment is excellent, and there are many examples of anodized aluminium that are giving good service around our coasts.

Another environment that can give rise to corrosion is the public swimming pool where aluminium is used for handrails, cubicles, etc. The atmosphere is humid and contains chlorine. Here it is recommended to increase the chemical resistance of the coating by sealing in a nickel salt solution (see page 94). A minimum coating thickness of 25 μm should be specified.

The resistance to attack by alkaline domestic detergents—a necessary requirement in washing machines—is again enhanced by nickel sealing the coating which will then give good service. Stronger alkaline solutions, such as are used in dish-washing machines, will attack the coating.

The coating is not resistant to attack by mineral acids and tends to be slowly attacked by the weaker organic acids.

Attack by acid or alkaline solutions is spread evenly over the surface and causes the coating to become opaque, whereas atmospheric attack generally results in the formation of small black pits sometimes with a white efflorescence at the pit opening. These pits can be rendered less conspicuous by cleaning off the deposits, but if they are not removed, the attack penetrates to the basis metal and beyond. Unlike many forms of corrosive attack on a coated metal the corrosive action does not spread sideways between the coating and the basis metal. Therefore there is no tendency for the coating to peel or flake off.

Chapter 15

Maintenance of Anodized Aluminium

Indoors the appearance of anodizing will be retained by periodic wiping with a soft dry cloth, possibly with occasional washing with water (which may contain soap or 1% Teepol), followed by drying and waxing with a good household polish (including most silicone types). Paint, grease and heavy dirt deposits may be removed by the use of a solvent cleaner, e.g. white spirit or carbon tetrachloride.

For exposure to outside atmospheric weathering thick anodic coatings may be kept in good appearance in most environments by washing every 1-12 months. Nylon brushes may be used to remove ingrained dirt. The greatest frequency of cleaning is necessary in combined industrial and marine atmospheres, and where the surface is not washed by rain. In rural atmospheres least cleaning is needed. Free access to falling rain is always beneficial. Thin (5 μm) coatings may be used out of doors only with very frequent cleaning; their life will be short and not comparable with the life required in many architectural applications. Car trim is a typical application of this type of thin film.

Surfaces that have grossly deteriorated and are severely pitted may have their appearance partially restored by careful use of fine stainless-steel wool, lubricated with soap or a light oil, followed by cleaning with pumice or metal polish. Mild-steel wool should not be used, as particles of it may remain behind and produce rust stains on the aluminium. The treatment leaves an impaired anodic film, so that lacquering or coating with wax or lanolin is needed to retain an attractive appearance. Such restored surfaces have lowered corrosion resistance, which may be partly offset by adding a wax emulsion to the water used for periodic cleaning.

Lacquering of anodized aluminium may also be advantageous on new work. It will help to resist damage in transportation and damage by contact with other materials, notably by mortars and plaster in building applications. Splashes of paint, cement, etc., may then be removed with dry fine steel wool. Proprietary lacquers available and recommended by the manufacturers for this purpose include some based on methacrylates and some based on cellulose acetate butyrate. The lacquer should be evenly applied, preferably by spraying: a thickness of 0.0006 in. is recommended. The total coating thickness, preferably applied as two coatings, should not be more than 0.001 in. Most of these lacquers will weather away in 2 to 4 years but can be renewed if required.

Transportation, storage and erection requires protection of the anodized surface in every case. As alternatives to clear, semi-permanent lacquer coatings, strippable lacquers or water-proof tapes may be used. These require removal at a late stage before the component goes into service, and many are not alkali resistant. A combination of an alkali-resistant wax emulsion followed by tape has also been used.

Chapter 16

Effluent Treatment for Anodizing Plants

This chapter outlines the effluent requirements that are likely to be encountered when operating an anodizing installation, and describes some of the methods that have been adopted to meet those requirements.

It must first be emphasized that effluent treatment is a specialist subject and that advice should be sought from supply companies having experience in this field. The question of effluent treatment is best considered at the stage when the anodizing plant location is under discussion. Some locations will involve meeting very stringent standards, for example where the effluent is to be discharged directly to a stream or river with intermediate treatment at a sewage works. In other locations a high existing flow rate of sewage will dilute the effect of an anodizing plant discharge and a large sewage works may be able to tolerate and treat a lower standard of discharge.

The local Water Authority must first be approached to ensure that they can accept the anticipated volume of effluent and to obtain from them a schedule of impurity limits that will be permitted. On the basis of this information a tentative plan for effluent treatment can be prepared for approval by the Water Authority.

Some of the major requirements that are likely to be enforced are as follows.

Volume

A restriction may be placed on the maximum volume of effluent that may be discharged in any one hour as well as on the total volume in a period of 24 hours. During normal production hours the effluent flow rate is usually uniform and predictable, but during cleaning-out operations, say at the week-end, there is a danger that the flow rate may be excessive. This will call for vigilance by the operating staff. Plants having a large holding tank for effluent can be organized so as to smooth out irregularities in the flow rate.

Temperature

A maximum temperature of 43°C is often cited. This is to protect the sewer maintenance staff from the peril of receiving a sudden surge of scalding liquid as well as to discourage too warm a temperature of the general sewage. When hot liquids such as hot-water sealing solutions have to be discharged they must be diluted with cold water from swill tanks.

pH

The pH limits vary considerably from area to area. Where the effluent is discharged directly to a stream or river a limit of 6-8 (in addition to other restrictions) is usual. In industrialized areas where the effluent will be diluted by large volumes of other sewage and where the whole will be treated in a sewage works a pH range of 6-10 is often acceptable.

The flow of acid rinse water from anodizing, desmutting and chemical brightening would normally mix with the alkaline rinse water from the alkaline cleaning and caustic soda etching. Where chemical brightening is in regular use the final rinse water mixture will be acidic and will require neutralizing which can be effected by dosing with

(a) caustic soda solution,
(b) lime slurry.

This process is easily automated. The choice of neutralizer depends upon other requirements — in particular the permissible level of sulphate (see "sulphate" section). If it is necessary to reduce the sulphate in the effluent the lime neutralizing process will cause calcium sulphate to be precipitated, but this precipitated sludge will, in turn, increase the solids content of the effluent (see "solid" section).

In some installations the caustic soda for neutralizing is provided by allowing spent caustic soda etching solution to stand for a few days at room temperature in a storage tank. During this time aluminium hydroxide is precipitated and caustic soda is regenerated. The aluminium hydroxide sludge will still need to be dealt with. An alternative source is a 50% solution of sodium hydroxide supplied in drums or by road tanker.

Effluent Treatment for Anodizing Plants

If the combined rinse waters produce an alkaline reaction outside the upper limit of pH neutralizing with acid is indicated. Diluted sulphuric acid, 50% v/v or weaker, is the cheapest source of acid and here again the dosing process can be automated. If the sulphate content of the effluent is already too high, the sulphuric acid must be replaced by hydrochloric acid—subject, however, to any limits on chloride called for in the effluent-discharge agreement.

The neutralizing processes described above cause the formation of a light gelatinous floc of aluminium hydroxide derived from the aluminium dissolved in the anodizing electrolyte, the desmutting solutions and caustic soda etches which contribute to the rinse-water composition.

Sulphate

The increasing use of concrete pipes for main sewers has resulted in a requirement to restrict the sulphate concentration in effluent to a maximum of 1200 mg/l as SO_4. Coincidentally, this figure is almost identical with the figure given by a saturated solution of calcium sulphate. Hence the lime-slurry-neutralizing process (see "pH" section) will achieve an acceptable residual sulphate figure.

However, in some localities a lower figure for sulphate is enforced which would not be provided by the lime process if applied to the final effluent. In this case the flow of rinse water from the sulphuric acid anodizers must be segregated and neutralized with lime followed by settlement or filtration to remove calcium sulphate sludge. The clarified liquor can then be mixed with the rest of the effluent stream which will dilute the residual sulphate to an acceptable figure.

Other methods of reducing the final sulphate figure are as follows:

(a) Increasing the effluent volume by using more water (an expensive exercise).
(b) Reducing the concentration of sulphuric acid in the anodizing electrolyte. Some increase in the operating voltage will be required to maintain the conventional current density of 1.2-1.5 amp/dm^2.
(c) Careful drainage of the anodized work over the anodizing solution before transferring to the rinse tank.

Chromium

Chromium in the hexavalent state (for example chromic acid and chromates) is highly toxic to the bacteriological system used by sewage works, and low limits such as 3 mg/l are imposed. In the reduced trivalent state (for example chromium sulphate) the toxicity is greatly reduced but, like many other heavy metals, its presence in sewage works sludge is detrimental to the use of the sludge on agricultural land.

Chromic acid and chromates arise from the rinse water after chromic acid anodizing, dichromate sealing and chromic acid desmutting solutions. They are best treated by providing a separate drain for all acid- and chromium-containing rinses which are then taken to a tank for automatic dosing with a reducing agent such as sodium metabisulphite solution which reduces the chromium to the trivalent state. When the acid mixture is finally neutralized the chromium will be precipitated as chromium hydroxide and treated with all the other solids produced at the neutralizing stage.

In small installations the reduction of chromium to the trivalent state can sometimes more conveniently be brought about by providing a static rinse after the chrome-containing solution followed by a running rinse. To the static rinse is added 5-10% commercial ferrous sulphate which immediately reduces the hexavalent chromium that is dragged into it. The static rinse is periodically emptied, dumped on an approved site and refilled with ferrous sulphate solution. This simplified system is not suitable for articles that will trap electrolyte or for items that are chromic anodized for crack detection.

Boron

Mention must be made of an occasional local requirement for a low boron figure where the sewage works effluent is later used for irrigating certain boron-sensitive crops or where the water will later be recycled for human consumption. Boron is contained as borate in some alkaline cleaners. The solution to this problem is to use a boron-free cleaner.

Nitrate

Whilst there is no generally agreed limit for nitrate in sewage its presence should be minimized where the treated sewage outfall is to be recycled later for potable water. A maximum figure of 50 mg/l in factory effluent has been suggested for the time being. This figure is unlikely to be reached unless nitric acid desmutting solutions are worked carelessly. Work should be drained over the desmutting tank before rinsing. If nitrate is causing serious problems the nitric acid type of desmutting system can be replaced in many instances by a ferric sulphate-based proprietary mixture with a very low nitrate content.

Suspended solids

One of the most vexatious problems in treating anodizing effluent is the removal of the suspended solids that are precipitated during the process of adjusting the pH of the effluent to about 7-9. The main product is a light flocculent gelatinous suspension of aluminium hydroxide together with the hydroxides of any other metals that have dissolved in the anodizing electrolyte, e.g. zinc, copper, manganese and magnesium, and metals present in metal-sealing solutions or electrolytic colouring baths, e.g. nickel, cobalt and tin.

If there is sufficient floor space available the hydroxide can be separated by gravity settlement in a large tank with a retention time of 2 hours or more. The clear overflowing liquor can then be discharged to the sewer or, in some cases, a portion — say 25% — can be recirculated as rinse water in the pre-anodizing section of the plant. The sludge contains 1-2% of solids and can be thickened by a slowly rotating paddle on the bottom of the tank. In this form it is expensive to dispose of by tankering away to a licensed dumping site.

The principle of air flotation can readily be applied to separating aluminium hydroxide. The plant occupies a relatively small floor area and gives a floc containing 2-3% aluminium hydroxide.

The "thin" sludge arising from the above two separation methods can be dewatered by one of three main methods:

(a) standing on a porous brick bed so that water drains off by gravity,
(b) pressure filtration,
(c) centrifuging.

Method (a) is slow and requires a large floor area. The thickened sludge has to be shovelled into skips for transport to a dump and is therefore costly in labour.

Pressure filtration is a well-established process which can be automated so that the pressed "cake" of sludge can fall into a skip for disposal. A solids content of 30-40% is obtainable.

Centrifuging is used in sewage works for concentrating sludge and is successfully applied to the separation of aluminium hydroxide. The best results will be achieved if the centrifuging is carried out on sludge that has already been somewhat concentrated by previous settlement or filtration. A cake containing about 15% solids is produced.

In all the separation processes mentioned above the rate and effectiveness of settlement or flotation can be improved by dosing the liquor beforehand with a polyelectrolyte.

The Water Authority requirements for suspended solids vary considerably according to the ultimate destination of the sewage. For example, where there is no intervening sewage works a maximum figure of 5 mg/l may be specified whilst in other areas with a large sewage works a figure of 500-1000 mg/l is quoted.

Organic acids

The organic acids most likely to be present in the effluent are oxalic, sulphophthalic or sulphosalicylic. Oxalic acid is toxic and is converted to insoluble calcium oxalate if lime dosing is available in the effluent-treatment plant. Oxalate is only present in very small concentrations arising from an addition of 1% oxalic acid to a sulphuric acid electrolyte. This practice is widespread in Europe and does not appear to constitute an effluent problem as far as rinse water is concerned, but the bath disposal needs special consideration.

Sulphophthalic and sulphosalicylic acids are the principal constituents of integral colour-anodizing solutions. As they are expensive chemicals to use it is customary to drain the work well before transferring from the anodizer to the rinse tank and the small amounts diluted into the effluent are usually acceptable.

Heavy metals

Installations that include nickel sealing or electrolytic colouring processes will contribute heavy metals such as nickel, cobalt, tin or copper or mixtures of them. If these remain in solution in the works effluent very strict limits are imposed on their single or combined concentrations, e.g. 2-10 mg/l.

Where a substantial quantity of aluminium hydroxide is precipitated at the neutralizing stage in the effluent-treatment plant there is a tendency for heavy metals to be absorbed by the hydroxide, but their presence in the precipitate means that the sludge cannot be used for agricultural purposes, but is generally suitable for land filling.

In view of the high market prices of the metals mentioned above the problem is better tackled by providing a static rinse tank after the metal-using process. The water in the tank is continuously circulated through a cationic resin column to absorb the metal which can subsequently be recovered by regenerating the column of resin.

Organic dyestuffs

The sudden arrival at a sewage works of coloured sewage can cause concern if no prior warning of this possibility has been given by the anodizer. Although the dyestuffs used by anodizers are not considered to be toxic their presence in sufficient quantity to leave a residual colour in the sewage-works effluent is clearly unacceptable. Such high concentrations can occur if the deeper shade solutions are dumped from large dye tanks. Depending on local requirements one of the following methods can be instituted in the anodizing works:

(a) Tankering away high-concentration dye solutions to an approved site.
(b) In the case of highly coloured rinse solutions, these can be circulated through activated carbon which absorbs the dye. The spent carbon can be incinerated or dumped.

Accidental spillage

However carefully an anodizing plant is operated there are rare occasions when a tank containing a noxious liquid can develop a serious and undetected leak. This may happen during a holiday period or over a week end. Furthermore, even a modern automated effluent-treatment plant can malfunction. One solution to this problem is to install a holding tank for the works effluent which will hold 1-2 hours' normal flow of effluent plus additional capacity to carry the contents of the largest tank of toxic material in the plant. This holding tank is provided with overflow outlet valves at two levels, (a) at the 1-2 hours' flow capacity level and (b) at the maximum capacity level.

During normal operations the lower valve is open. During holidays or in the event of any accidental spillage the lower valve is closed. Any special treatment of the liquid received into the holding tank can then be dealt with before re-opening the lower valve.

Appendix I

Approximate Equivalent Concentrations of Sulphuric Acid in Different Units

These figures are intended as a practical guide at 20°C (68°F). For exact equivalents and corrections for temperature BS 753: 1959 should be consulted.

TABLE 20 EQUIVALENT CONCENTRATIONS OF SULPHURIC ACID IN VARIOUS UNITS

% H_2SO_4 (w/w) (g in BS 753)	% H_2SO_4 (v/v)		Fluid ounces per gallon	H_2SO_4 grams/ litre (g in BS 753)	Normality (N)	Density* (D_{20} in BS 753)
8.7	5.0	(5.2)	6.7	92	1.87	1.057
9.3	5.3	(5.6)	7.1	98	**2.0**	1.061
10.0	5.7	(6.0)	7.6	106	2.18	1.066
11.1	**6.5**	(6.8)	8.7	120	2.43	1.074
12.0	7.0	(7.3)	9.3	129	2.66	1.080
13.5	8.0	(8.4)	10.7	147	**3.0**	1.091
13.7	8.2	(8.6)	10.9	**150**	3.06	1.093
15.0	9.0	(9.5)	12.0	165	3.37	1.102
15.8	9.5	(10.0)	12.7	174	3.57	1.108
16.5	**10.0**	(10.5)	13.3	183	3.75	1.113
17.8	10.9	(11.5)	14.6	**200**	4.08	1.123
20.00	12.4	(13.0)	16.6	228	4.65	1.139
21.2	13.3	(14.0)	17.7	245	**5.0**	1.149
21.7	13.6	(14.3)	18.1	**250**	5.10	1.153
23.6	15.0	(15.8)	20.0	275	5.63	1.167
30.2	**20.0**	(21.0)	26.7	368	7.51	1.220
32.6	22.0	(23.2)	29.3	404	8.28	1.240

*Only applicable to new baths.

Appendix I 151

As supplied to the trade, concentrated sulphuric acid is usually of about 95% concentration. While most of the table refers to pure acid, the second column includes figures in brackets which refer to the percentage required of the usually available acid concentration. Bold figures are some of those frequently quoted in the literature.

Appendix II

Bath Analysis Methods

CAUSTIC SODA ETCHING SOLUTION

The modern etching formulations contain sequestering agents which interfere with the accurate determination of the caustic soda and aluminium figures. However, a high degree of accuracy is unnecessary for commercial requirements and the following procedure will usually suffice:

Filter a sample of the warm etch solution through a coarse paper. Allow the filtrate to cool. Pipette 10 ml of the filtrate into a 250-ml conical flask. Add 50 ml distilled water. Titrate the solution with 0.5 M sulphuric acid to a faint cloudiness or to an end point with thymolphthalein indicator. Record the volume of sulphuric acid used (A ml).

Add a few drops of phenolphthalein indicator and continue the titration to a colourless end point. Record the additional volume of acid used (B ml).

$$A \times 4 = \text{g/l sodium hydroxide,}$$
$$B \times 2.2 = \text{g/l aluminium.}$$

NITRIC ACID DESMUTTING SOLUTION

Pipette 25 ml of the solution into a 250-ml volumetric flask and make up to the mark with distilled water and mix well. Pipette 25 ml of the diluted solution into a 250-ml conical flask. Add 1 g neutral potassium fluoride and dissolve. Add a few drops of thymol blue indicator solution and titrate with M sodium hydroxide to the blue end point. Record the volume used (A ml).

Nitric acid (% v/v concentrated acid d. 1.42) = $A \times 2.52$.

NITRIC ACID AND ALUMINIUM IN PHOSPHORIC/NITRIC TYPE CHEMICAL POLISHING SOLUTIONS*

Nitric Acid

Reagents. All should be of A.R. quality.

1. *Ferrous sulphate solution.* Cautiously add 50 ml conc. sulphuric acid to about 700 ml distilled water. Cool, add approximately 250 g ferrous sulphate heptahydrate ($FeSO_4.7H_2O$) and stir until dissolved. Make up to 1 litre, mix well and transfer to a clean dry bottle with a well-fitting stopper. Standardize monthly.
2. *Phosphoric acid* (s.g. 1.75).
3. *Standard phosphoric-nitric acid solution.* Using a burette or a safety pipette, add 25 ml 60% nitric acid to a 500-ml volumetric flask and make up to the mark with phosphoric acid (s.g. 1.75). Mix well, keep stirred and store in a cool place. Make up freshly every 3 months.

Standardization of ferrous sulphate solution

Accurately measure 5.0 ml standard phosphoric-nitric acid solution into a dry 250-ml beaker: a conical flask must not be used for this titration. A burette with a wide aperture is the most convenient way of measuring the viscous solution. Add 100 ml phosphoric acid. Stir and heat to 40-45°C. Titrate with ferrous sulphate solution until the first permanent golden-brown colour is formed. Towards the end of the titration there will be numerous gas bubbles dispersed throughout the solution and reddish-brown fumes are evolved. At the end point no fumes are evolved. Let A ml be the volume of ferrous sulphate used.

Procedure

Measure 5.0 ml brightening solution into a clean dry 250-ml beaker, and proceed as described in the previous section, titrating

---

*Method supplied by Albright & Wilson Ltd., the proprietors of the Phosbrite range of chemical brighteners.

against ferrous sulphate. Let B ml be the volume of ferrous sulphate used.

Calculation

% by volume of 60% nitric acid in the brightening solution is

$$\frac{B}{A} \times 5.$$

Aluminium

Principle of the method

An aliquot of diluted brightening solution is taken and buffered and EDTA added in excess to produce the aluminium-EDTA complex. The excess of uncomplexed EDTA is taken up by titrating with copper sulphate solution. Sodium fluoride is then added in a sufficient quantity to decompose the aluminium-EDTA complex and the released EDTA is further titrated with standard copper sulphate. This last titration is a measure of the aluminium concentration.

Reagents. All should be of A.R. quality.
1. *Copper sulphate solution. 0.1 molar.* Dissolve 24.97 g copper sulphate pentahydrate in distilled water; add one or two spots of concentrated sulphuric acid, transfer to a 1-litre volumetric flask and dilute to the mark with distilled water. Mix well.
2. *EDTA solution. 0.1 molar.* Dissolve 37.2 g EDTA disodium salt dihydrate in distilled water. Transfer to a 1-litre volumetric flask and dilute to the mark with distilled water. Mix well.
3. *PAN indicator solution. 0.1% w/v.* Dissolve 0.1 g 1-(2-pyridylazo)-2-naphthol (PAN) in 100 ml industrial methylated spirits.
4. *Buffer solution.* Dissolve 500 g ammonium acetate in distilled water; add 20 ml glacial acetic acid and dilute to 1 litre with distilled water.
5. *Sodium fluoride.*

Procedure

Measure 5.0 ml brightening solution from a burette into a 100-ml standard flask, allowing time for complete drainage. Dilute almost to the mark with distilled water and mix thoroughly. Cool and top up to the mark and again mix.

Pipette 10 ml of the prepared solution into a 600-ml beaker and dilute to 350-400 ml. Add 20 ml buffer solution and 25 ml 0.1 M EDTA. Warm and add 1.0 ml PAN indicator. Bring to the boil, cool to 60°C and titrate with standardized 0.1 M copper sulphate solution. The end point is shown by a sharp change from green to deep blue.

Add 1-2 g sodium fluoride to decompose the aluminium-EDTA complex and again bring to the boil. Cool to 60°C and titrate the liberated EDTA with 0.1 M copper sulphate to the blue end point. Call this titration "z" ml. Then "z" × 5.4 = g/l aluminium.

SULPHURIC ACID ANODIZING SOLUTION

Free sulphuric acid

Dilute 50 ml of the electrolyte to 250 ml in a volumetric flask. Pipette 10 ml of the diluted solution into a 250-ml conical flask and add 20 ml distilled water, 1 g potassium fluoride and 1 ml Thymol Blue indicator solution. Titrate with M sodium hydroxide to a blue end point. Record the volume used (A ml).

$$\text{Sulphuric acid (\% v/v)} = 1.33A$$
$$\text{(of commercial acid d. 1.84).}$$

Aluminium

Pipette 25 ml of the electrolyte, diluted as for the free acid determination, into a 250-ml conical flask and add an excess (B ml) of M sodium hydroxide and 1 ml Thymol Blue indicator solution. Titrate with M nitric acid to a yellow end point. Record the volume of nitric acid used (C ml).

$$\text{Aluminium (g/l)} = 1.8(B-C) - 4.5A.$$

This method gives results that are about 5% lower than those given by the atomic absorption or gravimetric methods.

Chloride

To 100 ml of electrolyte add 10 ml nitric acid (d. 1.42); then add 50 ml of 1% silver nitrate and boil the solution; when the precipitate has settled, filter through a weighed Gooch crucible, wash with hot 2% nitric acid, dry at 110°C, cool and weigh.

$$\text{NaCl (g/l)} = \text{g AgCl} \times 4.1.$$

SULPHURIC PLUS OXALIC ACID ANODIZING SOLUTIONS*

Take 10 ml from the bath and dilute to 100 ml.

Total oxalic acid

Place 10 ml of the diluted solution (corresponding to 1 ml of the original bath) in a 250-ml conical flask. Dilute to 70 ml and add 20 ml of 50% H_2SO_4.

Heat to 60-70°C and titrate with 0.04 M potassium permanganate until a permanent pink coloration is obtained. Let the 0.04 M potassium permanganate used be x ml. Then the oxalic acid content is:

$$\text{Oxalic acid (g/l)} = 4.5x.$$

Total acidity

The total acidity is the sum of the sulphuric and oxalic acids free and combined with aluminium.

Titrate 10 ml of the diluted solution with a solution of 0.2 M sodium hydroxide, using an alcoholic solution of phenolphthalein as indicator.

Let the sodium hydroxide solution be y ml. Then, in terms of sulphuric acid:

$$\text{Total acidity (g/l } H_2SO_4) = 9.8y.$$

Total sulphuric acid

The total weight of sulphuric acid (free and combined) is:

$$\text{Sulphuric acid (g/l)} = 9.8y - x/2.$$

Free acid

The free acid is the amount of oxalic acid and sulphuric acid not combined with aluminium.

Take 10 ml of the diluted solution. Add 20 ml of a 50% solution of **neutral potassium fluoride. Titrate with a 0.2 M solution of sodium** hydroxide, using bromothymol blue as indicator. Let z be millilitres of sodium hydroxide solution used in the titration. Then weight of free acid is:

$$\text{Free acid (g/l } H_2SO_4) = 9.8z.$$

*Translated from *Revue de l'Aluminium*, **34** (241), 297-8 (1957).

Free sulphuric acid

The free sulphuric acid is:

$$\text{Free sulphuric acid (g/l)} = 9.8(z-x/2).$$

Aluminium

The aluminium content of the bath is given by the formula

$$\text{Aluminium (g/l)} = 1.8(y-z).$$

CHROMIC ACID ANODIZING SOLUTIONS*

Total chromium as CrO_3

Dilute 25 ml of the electrolyte to 250 ml in a volumetric flask. Pipette 25 ml of the diluted solution into a 500-ml conical flask containing 150 ml distilled water and 45 ml of 25% v/v sulphuric acid. Add 10 ml 3% silver nitrate solution, 2 g ammonium persulphate and boil for 20 minutes. Cool to room temperature, add 5 drops of N-phenylanthranilic acid and titrate with 0.1 M ferrous ammonium sulphate until the reddish-purple colour changes to green. Allow a few seconds to elapse after each drop of titrant towards the end point. Record the volume of ferrous ammonium sulphate (B ml).

Total chromium as

$$CrO_3 = \frac{125}{3A} \times B \text{ g/l}$$

where A is the "factor" for the ferrous ammonium sulphate solution when titrated as above against 25 ml of M/60 potassium dichromate solution instead of the electrolyte. If A ml of ferrous ammonium sulphate are required the "factor" = $125/A$.

Free chromic acid as CrO_3

Used electrolyte

Pipette a 10-ml sample of the electrolyte into a 250-ml conical flask containing 100 ml deionized water and titrate with 0.25 M sodium

*Taken from Defence Specification DEF 151.

carbonate solution to the first appearance of a distinct permanent turbidity. Record the volume of sodium carbonate required (C ml).

Calculation

$$\text{Free chromic acid as } CrO_3 = 5C \text{ g/l.}$$

Fresh electrolyte

Dilute a 25-ml sample of the electrolyte to 250 ml in a graduated flask. Pipette 50 ml of this solution into a 500-ml conical flask containing 45 ml 25% v/v sulphuric acid and 200 ml distilled water. Add 5 drops of N-phenylanthranilic acid indicator and titrate with 0.1 N FAS. Record the volume of FAS used.

Calculation

$$\text{Free chromic acid as } CrO_3 = \frac{50}{3} \times \frac{D}{A} \text{ g/l.}$$

Determination of sulphate content

To 100 ml of filtered electrolyte in a 400-ml squat-form beaker add 10 ml conc. hydrochloric acid, 25 ml glacial acetic acid and 20 ml ethanol. Boil gently for 15 minutes to expel aldehydes and excess ethanol. Dilute to 200, bring back to the boil and, while boiling, add 10 ml of 10% w/v barium chloride solution. Continue to boil for a further 15 minutes and digest for a total of 1 hour; if possible allow to settle overnight. Filter the precipitate on a weighed No. 4 Gooch crucible and wash with hot water; dry, cool and re-weigh. Calculate weight of barium sulphate (E g).

Calculation

$$\text{Sulphate as } Na_2SO_4 = 6.1E \text{ g/l.}$$

Determination of chloride content

Transfer 100 ml of electrolyte to a 300-ml flask and add 10 ml of concentrated nitric acid. Heat to boiling, add 50 ml 1% w/v silver nitrate solution and agitate vigorously to coagulate the precipitate.

Filter on a weighed No. 4 Gooch crucible, washing with hot, dilute (2% v/v) nitric acid. Dry, cool and re-weigh the crucible. Record the weight of silver chloride (F g).

Calculation

$$\text{Chloride as NaCl g/l} = 4.1F.$$

FERRIC AMMONIUM OXALATE DYE SOLUTIONS

Determination of iron

Pipette 20 ml of the filtered solution into a 250-ml beaker. Add 20 ml purified water and 10 ml 20% v/v sulphuric acid. Heat to 50-60°C and titrate with 0.04 M potassium permanganate, stirring constantly until a persistent pale-pink colour appears. Adjust the pH to 2 by adding about 20 ml of a buffer solution made up from 164 g anhydrous sodium acetate and 100 g chloracetic acid crystals per litre. Add 1 ml of indicator solution (200 g of 5-sulphosalicylic acid dihydrate per litre), producing a deep-red colour. Titrate with 0.1 M EDTA at 60°C until the colour becomes pale yellow. Record the volume of EDTA (A ml).

$$\text{Ferric ammonium oxalate} = 3.09 \times A \text{ g/l}.$$

COBALT ACETATE/POTASSIUM PERMANGANATE PIGMENT COLOURING SOLUTIONS

Cobalt acetate

Pipette 10 ml of solution into a 300-ml beaker. Add distilled water to about 200 ml. Add 2 g ammonium chloride, 5 g sodium acetate crystals and a few drops of Murexide indicator. Add ammonia with constant stirring until the orange-red colour becomes gold. Titrate with 0.1 M Komplexon III to a red-violet end point. Record the volume used (A ml).

$$\text{Cobalt acetate} = 2.49 \times A \text{ g/l}.$$

Potassium permanganate

A number of classical methods are available for this determination. The following is simple and straightforward.

Pipette 20 ml of 0.05 M oxalic acid solution into a 300-ml beaker. Dilute with about 150 ml distilled water and add 20 ml 50% v/v sulphuric acid. Warm to a temperature of 60-70°C. Titrate with the potassium permanganate solution until a faint permanent pink colour appears. Record the volume of potassium permanganate used (A ml).

$$\text{Potassium permanganate} = 63.22 \times A \text{ g/l.}$$

Appendix III

Selected Books — Information Sources

The principal literature on the subject of anodizing aluminium dates from 1924 and in view of the adoption of the process on a world-wide basis the technical and patent literature on the subject is considerable.

For those wishing to learn more about this the most interesting of metal-finishing processes, it is recommended that the following books should be consulted:

Anodic Oxidation of Aluminium and its Alloys, Jenny. (Translated by W. Lewis.) Charles Griffiths & Co. Ltd., London. 231 pp.

Werkstoffe Aluminium und seine anodische Oxydation, M. Schenk. A. Francke AG Verlag, Berne, 1948. 1042 pp.

An English translation of this book is not available. It covers very fully the state of the art at the time of publication and carries references to most of the important sources at that time.

The Surface Treatment and Finishing of Aluminium and its Alloys. S. Wermilk and R. Pinner. Robert Draper Ltd., Teddington. Fourth edition 1972. 2 volumes. 1274 pp.

This is a worthy successor to Schenk's book and is regarded as the "encyclopaedia" on the subject.

Die Praxis der anodischen Oxydation des Aluminiums. Dr. W. Hübner. C. T. Speiser Aluminium - Verlag GmbH, Düsseldorf. Third edition 1977. 493 pp.

An English translation of this edition is believed to be in preparation. It is a good practical book reflecting Continental practices. It lists many of the more important patents in this field.

The Technology of Anodizing Aluminium. A. W. Brace, P. G. Sheasby. Technicopy Ltd., Stonehouse. 2nd edition 1979. 321 pp.

Again we have a practical book with useful guidance on the economics of the anodizing business. It has a large list of literature references up to 1978.

In addition to the above, the following U.K. organizations provide advice and information on the subject:

Aluminium Federation Ltd.,
  Broadway House, Calthorpe Road, Five Ways, Birmingham B15 1TN.
BNF Metals Technology Centre,
  Denchworth Road, Wantage, Oxfordshire, OX12 9BJ.
British Aluminium Co. Ltd.,
  Chalfont Park, Gerrards Cross, Buckinghamshire.
Alcan International Ltd.,
  Banbury Laboratories, Southam Road, Banbury, Oxfordshire.
Alcoa of Great Britain Ltd.,
  Alcoa House, PO Box 15, Droitwich, Worcestershire, WR9 7BG.

The following U.K. technical journals publish papers dealing with anodizing and their indexes should be consulted:

Transactions of the Institute of Metal Finishing, Institute of Metal Finishing, Exeter House, 48 Holloway Head, Birmingham, B1 1NQ.
Product Finishing, Sawell Publications Ltd., 127 Stanstead Road, Forest Hill, London, SE23.
Finishing Industries, Wheatland Journals Ltd., 177 Hagden Lane, Watford, WD1 8LW.

References are made in the text to many of the currently available aluminium alloys. Full details of their chemical compositions, mechanical properties and temper designations appear in a booklet *The Properties of Aluminium and its Alloys* issued by the Aluminium Federation Ltd. at the address given above.

Appendix IV

Specifications Applicable to Anodic Oxide Coatings on Aluminium

Most of the industrialized countries have published process or testing specifications for anodizing. A prime source of information on this subject is the British Standards Institution, 2 Park Street, London W1A 2BS, from which copies of British Standards and of most other national and international standards can be purchased.

The following is a list of British Standards (BS) and International Standards (ISO) that are relevant to the anodizing industry. It must be noted that the principal BS Specification in this field, 1615, is about to be replaced by a series of standards comprising a head specification listing test requirements and a series of test method standards.

In the list of International Standards the prefix DIS denotes that the document is in draft form and likely to be adopted with little modification.

U.K. STANDARDS (BS)

BS 1615 Anodic Oxidation Coatings on Aluminium
BS 3987 Anodic Oxide Coatings on Wrought Aluminium for External Architectural Applications
BS 5599 Specification and Methods of Test for Hard Anodic Oxide Coatings on Aluminium for Engineering Applications
BS AU 89 Anodized Aluminium for Automobile Use

EUROPEAN STANDARDS

Apart from the national standards of the European countries the

European Anodizing Association (EURAS) has developed and sponsored the "Qualanod" scheme for the licensing of anodizing installations that comply with the requirements of the "Qualanod" specifications. U.K. inquiries should be sent to

Qualanod (Great Britain) Ltd.,
Aluminium Federation Ltd.,
Broadway House, Calthorpe Road,
Five Ways, Birmingham B15 1TN.

INTERNATIONAL STANDARDS (ISO)

ISO 1463	Metallic and oxide coatings—Measurement of thickness by microscopical examination of cross-sections.
ISO 2064	Metallic and other non-organic coatings—Definitions and conventions concerning the measurement of thickness.
ISO 2085	Surface treatment of metals—Anodizing of aluminium and its alloys—Check of continuity of thin coatings—Copper sulphate test.
ISO 2106	Surface treatment of metals—Anodizing (anodic oxidation) of aluminium and its alloys—Measurement of the mass of the oxide coatings—Gravimetric method.
ISO/R 2128	Surface treatment of metals—Anodizing (anodic oxidation) of aluminium and its alloys—Measurement of thickness of oxide coatings—Non-destructive measurement by split beam.
ISO 2135	Anodizing of aluminium and its alloys—Accelerated test of lightfastness of coloured anodic oxide coatings.
ISO 2143	Surface treatment of metals—Anodizing of aluminium and its alloys—Estimation of the loss of absorptive power by colorant drop test with prior acid treatment.
ISO 2360	Non-conductive coatings on non-magnetic basis metals—Measurement of coating thickness—Eddy current method.
ISO 2376	Anodizing (anodic oxidation) of aluminium and its alloys—Insulation check by measurement of breakdown potential.

Appendix IV

ISO 2767	Surface treatment of metals — Anodic oxidation of aluminium and its alloys — Specular reflectance at 45° — Total reflectance — Image clarity.
ISO 2813	Paints and varnishes — Measurement of specular gloss of non-metallic paint films.
ISO 2859	Sampling procedures and tables for inspection by attributes.
ISO 2931	Anodizing of aluminium and its alloys — assessment of quality of sealed anodic oxide coatings by measurement of admittance or impedance.
ISO 2932	Anodizing of aluminium and its alloys — Assessment of sealing quality by measurement of the loss of mass after immersion in acid solution.
ISO 3210	Anodizing of aluminium and its alloys — Assessment of sealing quality by measurement of the loss of mass after immersion in phosphoric-chromic acid solution.
ISO 3211	Anodizing of aluminium and its alloys — Assessment of resistance of anodic coatings to cracking by deformation.
ISO 3770	Metallic coatings — Copper accelerated acetic acid salt-spray test (CASS test)
ISO 3843	Anodizing of aluminium and its alloys — Accelerated test of lightfastness of coloured anodic oxide coatings.
ISO 6581	Anodizing of aluminium and its alloys — Fastness to ultraviolet light.
ISO/DIS 6719	Anodizing of aluminium and its alloys — Measurement and calculation of reflectance characteristics of aluminium surfaces using integrating sphere instruments.

MINISTRY OF DEFENCE SPECIFICATION

DEF - 151   Anodizing of Aluminium and Aluminium Alloys. Published by H.M. Stationery Office.

Index

Abrasion resistance 119-21, 130
Abrasive jet test 119, 121, 130
Abrasive wheel test 119, 122
Acetic acid salt spray test 112
Acetic acid/sodium acetate test 111
Acid etching 50
Acidified sulphite test 110
Activating for dyeing 77
Adhesive bonding 18, 27
Admittance test 108
Ageing of coatings 109
Agitation of solutions 32, 57, 58
Alkaline cleaner 49
Alkaline etching 50
"Alumilite" process 56, 68
Aluminium
   bright trim material 13
   cast anodizing characteristics 13, 15
   extrusions 14, 15
   forgings 15
   grades for anodizing 11
   sheet 14
   wrought anodizing characteristics 12
Aluminium Blue G 78
Aluminium Blue 2LW 108
Aluminium in chromic acid 71-72
Aluminium Deep Black MLW 78
Aluminium Red B3LW 78, 107, 108
Aluminium sulphate, effect of 59, 63
Aluminium Turquoise PLW 78
Aluminium Yellow 3GL 78
"Aluprint" process 83
"Alzak" process 9, 35, 53
Analysis of solutions 152-60
Anodic oxide coatings
   abrasion resistance 130
   admittance 108
   appearance 18

colour 20, 100
corrosion resistance 21, 139
density 103, 124, 126, 127
dimensional change 20
effect of anodizing conditions 127
electrical insulation 21
hardness 129
heat radiation 21, 136
heat reflection 21, 136, 137
heat resistance 135
isolation of 98
light fastness 134
permittivity 137
power factor loss 138
refractive index 134
structure 4, 125
testing 100-23
thickness 102, 124
Anodized aluminium applications
   anti-marking 8
   basis for organic coating 6, 63, 66, 89
   coloured 7
   corrosion resistance 6
   electrical insulation 10
   heat reflection 9
   lighting equipment 9
   lubrication 9
   wear resistance 9, 121
Anodizing
   hollow components 45
   sheet 47
   small components 47, 48
   strip 63, 65
   wire 64, 65
Anodizing aluminium
   agitation 32
   ancillary services 31

Index

Anodizing aluminium (*cont.*)
  drainage 30
  electricity 30
  equipment for 29-41
  gas 31
  premises 29
  steam 31
  water 30
"Anolok" process 35, 81
"Asada" process 81
Automatic anodizing 39

Banding finish 26
Barrier layer coatings 4, 74, 124, 138
Beer barrel anodizing 89
Bend test 129
Bending large sections 25
Bengough-Stuart process *see* Anodizing processes, chromic acid
Beta-backscatter test 105
Bleaching colours 83, 98
Bloom on anodic coatings 57, 93, 109, 111
Blueprint process 85
Bright anodizing 13, 14, 17, 21, 25, 132, 134
Brightening
  chemical *see* "Phosbrite"
  electrolytic 52, 53
British Standards 163
"Brytal" process 9, 35, 53, 133

Calibration of anodizing conditions 59
CASS test 113
Cation exchange for chromic acid 72
Caustic soda etching 50, 152
Chemical brightening *see* "Phosbrite"
Chemical cleaning 49
Chemical sealing *see* Sealing processes
Chloride
  in boric acid 75
  in chromic acid 71
  in sulphuric acid 63
Chrome/phosphoric desmutting 54
Chrome/phosphoric sealing test 111
Chrome/phosphoric stripping 97
Chromic acid anodizing *see* Anodizing processes

Cleaning processes 49
Coating weight
  effect of concentration 61
  effect of temperature 60
Cobalt/permanganate colour 8, 80, 159
Colouring
  electrolytic *see* Electrolytic colouring
  for identification 8
  integral *see* Integral colours
Colouring processes 76-87
Compressed air equipment 32
Continuity of coating test 121
Continuous anodizing 63-66
Cooling systems 33
Corrosion resistance tests 112, 133
Crack detection 15, 18, 71
Crazing of coating 66, 122, 128, 129, 138

Designing for anodizing 23
Desmutting
  in chromic acid 54, 55
  in ferric sulphate 55
  in nitric acid 55, 153
  in phosphoric/chromic acid 54
Dichromate-albumin process 86
Dichromate-gelatine 86
Dichromate sealing *see* Sealing processes, dichromate
Diffuse reflectivity 116, 132
Dimensional changes 66
Drawing, forming and spinning 24
Dye spot test 107
Dyed coatings, effect of heat 78
Dyestuffs for colouring 78
Dynamo for DC supply 35

Eddy current thickness test 104
Electrical breakdown test 122
Electrolysis, principle 3
Electrolytic brightening 52
Electrolytic colouring 20, 81, 119, 135
Electrolytic polishing 51, 52
"Eloxal" processes 56, 73
Emulsion cleaner 49
Etching
  alkaline 50, 83
  caustic soda 50, 83

Etching acid   50, 51, 83
"Eurocolor" process   81
European Standards   163

Fatigue strength   67, 68, 128
Ferric ammonium oxalate gold dye
   8, 80, 135, 159
Ferrous sulphate, chrome stain
   removal   71
Filler and electrode wires for
   anodizing   24
Flexible coatings   65
Fluoride in sulphuric acid   63
Friction between coatings   131
Fume extraction   35, 36, 71

Gardam grid test   116
Gravimetric coat thickness test   103
Grinding   24

Hard anodizing   17, 19, 21, 34,
   66-68, 129
"Hardas" process   67
Hardness of coating   129
Heat conservation   31
Heat exchangers   34
"Hiduran" process   67

Image clarity test   116
Information sources   161-2
Infra-red reflectivity   21, 117, 136, 137
Inorganic pigments   80
Integral colours   8, 14, 19, 20, 21, 32,
   119, 135
International (ISO) Standards   164, 165

Jigging (racking) for anodizing   42-48

Kape test   110

Lacquer on anodic coatings   6, 63, 66,
   88, 89, 138
Lightfastness   77, 78, 118

Maintenance of anodized aluminium
   141
"Martin Hard Coat" process   68
Mechanical agitation   32
Mechanical pretreatment   23
"Metacolor" process   81

Micron, definition   1
Microscopic section for coating
   thickness   102
Multicolour processes   8, 81-83

Neolan dyes   8
Nickel acetate sealing   see Sealing
   processes
Nigrosine D dye   8
Nitric acid desmutting   see
   Desmutting

Organic dyes for colouring   79
Oxalic acid anodizing   see
   Anodizing processes
Oxide film, natural   1

"Phosbrite" process   54, 55, 133,
   134, 153
Phosphoric acid
   anodizing   see Anodizing processes
   contamination by   63
Photographic processes   84-87
Pigment colours   20, 80, 135, 159
Plant layout   39
Properties   see Anodized aluminium
   applications
PRS head for reflectivity test   114

"Qualanod" licensing scheme   164

Rack stripping   46
Racking (jigging) for anodizing   42-48
Rectifiers   34, 57
Reflectivity tests   113-17
Reflectors
   floodlight   9
   infra-red   9
   specular   131-3
Refrigeration   33
Resin bonding   27
Rinsing   41
Riveting   27

Salt spray test   113
Sandblasting   23
"Sanford" process   68
Schuh and Kern test   119, 120, 121
Scott test   107
Scratch brushing   26

Sealing, quality testing 107-12
Sealing processes
  anodized wire 65
  chemical 89
  cobalt acetate 94
  continuous strip 65
  dichromate 6, 68, 95, 96, 128
  "Honnylite" 88
  hot water 7, 65, 71, 89-93
  lacquer *see* Lacquer on anodic coatings
  lanolin 7, 71, 88
  lubricating oil 9
  nickel acetate 84, 94, 95
  paint 6
  physical 88
  PTFE 10, 88
  steam 7, 89-91
Solway Blue BS 8
Solway Ultra Blue 8
Specifications for anodizing 163-5
Specular reflectivity test 114
Split beam microscope 102, 103, 104
Spot welding 27
Stopping off 46, 47
Stripping coatings 97-99
Stripping racks 46
Sulphophthalic acid process 74
Sulphosalicylic acid process 74
Sulphur-dioxide/humidity test 109
Sulphuric acid
  anodizing *see* Anodizing processes
  cleaner 150
  concentration tables 150

Tank construction 36, 37
Temperature, effect on coating 60
Temperature control 62
Thickness of coating
  calculation 60
  chromic acid 69
  measurement 102-7
Throwing power 62
Titanium for racks 42-44
Total reflectivity test 114
Trichlorethylene degreasing 49

UVIARC test 118

Vapour blasting 23
Voltage
  chromic acid process 70
  effect on current density 58
Voltage breakdown 138

Water
  impurities 39
  purification 38
Welding processes 25
Welding stain 25
Welding wire 26
Wire, anodized 10, 64, 65